NeuroDynamix

NeuroDynamix

Computer Models for Neurophysiology

W. Otto Friesen

Jonathon A. Friesen

New York Oxford
OXFORD UNIVERSITY PRESS
1994

Oxford University Press

Oxford New York Toronto
Delhi Bombay Calcutta Madras Karachi
Kuala Lumpur Singapore Hong Kong Tokyo
Nairobi Dar es Salaam Cape Town
Melbourne Auckland Madrid

and associated companies in
Berlin Ibadan

Published by Oxford University Press, Inc.,
200 Madison Avenue, New York, New York 10016

Library of Congress Cataloging-in-Publication Data
Friesen, W. Otto, 1942–
NeuroDynamix: computer models for neurophysiology/
developed by W. Otto Friesen and Jonathon A. Friesen.
p. cm. Includes bibliographical references and index.
ISBN 0-19-508282-6
1. Action potentials (Electrophysiology)—Computer simulation.
2. Neural conduction—Computer simulation.
3. Neurons—Computer simulation.
I. Friesen, Jonathon A., 1973–
II. Title
QP363.F75 1994 591.1'88—dc20 94-13940

9 8 7 6 5 4 3 2 1
Printed in the United States of America
on acid-free paper

To Family and Friends

PREFACE

OBJECTIVES

The transmembrane electrical potential is a critical aspect of the function of all living cells. In neurons, the dynamics of this potential are of particular importance as the basis for cell signaling. Although the electrical properties of cells are addressed in cell biology courses, bioelectricity is usually covered in greatest depth in courses on neurobiology. Due to the difficulty of electrical concepts and the static nature of textbook illustrations, students usually find the topic of bioelectricity both intimidating and inaccessible. Our primary objective in developing the *NeuroDynamix* modeling system was to overcome these twin problems. The modeling system provides a means for students to have access to neurophysiological methods and results by carrying out experiments. Such hands-on simulations serve to deepen students' understanding of electrophysiological concepts and to heighten their appreciation of the techniques used to study the electrical properties of cells. Unlike the static illustrations found in textbooks, the models of *Neurodynamix* simulate physiological experiments and results dynamically; results are displayed as they are being generated by the models. Based on a user-friendly, highly accessible graphics interface, the computer models encourage active exploration of physiological properties through the manipulation of model parameters while the model experiments are in progress.

ORGANIZATION

The modeling system has two interdependent components—this text and the accompanying software. Section I of the text provides overviews of electrical concepts, the properties of ion channels, resting and action potentials, synaptic interactions, and neuronal circuits. Each of these is followed by descriptions of modeling exercises to be carried out with the *NeuroDynamix* software. Exercises are designed to illustrate the concepts introduced in the didactic overviews. Grid lines for recording data from the computer screen or for producing graphs are included with the text as an aid to the user. Section II provides brief descriptions of the six models incorporated into *NeuroDynamix* and includes glossaries for variables and parameters. Sections III and IV furnish detailed descriptions of the mathematical equations for the models. Finally, the appendix includes detailed instructions for using the windows that are central features of this program. These instructions are also available on-line during execution of the *NeuroDynamix* program.

MAJOR FEATURES

NeuroDynamix simulates the dynamic properties of neurons at four levels of neuronal organization: the membrane patch, neuronal compartments, individual neurons, and neuronal circuits. Each model is designed to reflect its specific organizational level. Thus biophysical properties of individual ionic channels are modeled in detail by PATCH, but only the macroscopic behavior of an ensemble of channels is modeled at the higher organizational levels simulated in SOMA, AXON, NEURON, and CIRCUIT. There are two reasons for this approach to neuronal modeling. First, by modeling macroscopic rather than microscopic events at high levels of organization, parameters appropriate for that level can be incorporated into the models. Second, program execution is much faster if vast numbers of microscopic events are combined into a few macroscopic equations. Thus even on a computer with moderate computational speed the variables are graphed *dynamically*.

SUGGESTIONS FOR USE

These computer models were developed as a supplement for undergraduate and graduate courses in Cellular and Molecular Neurobiology taught in the Biology Department at the University of Virginia. However, this material also provides an excellent program for self-study, appropriate for advanced undergraduate students or graduate students who wish to master electrophysiology. It is suitable as well as a supplement for upper-level undergraduate or graduate cell biology courses that include an electrophysiological component. The modeling exercises may be employed either in a group setting or by individuals studying on their own.

Although not intended as a substitute for a "wet" neurophysiology laboratory, the computer models can be used to illustrate many of the principles demonstrated in such laboratories. Properly orchestrated manipulations performed with *NeuroDynamix* can simulate a laboratory setting without the expense, instructional effort, and student frustration that inevitably accompany student laboratories in electrophysiology.

W. OTTO FRIESEN
Department of Biology
Center for Biological Timing
University of Virginia
Charlottesville, VA 22903

JONATHON A. FRIESEN
2821 Northfields Road
Charlottesville, VA 22901

CONTENTS

ACKNOWLEDGMENTS

We would like to acknowledge the assistance and encouragement of the numerous women and men who assisted with the development of this book and computer program. We thank the students in the Friesen lab at the University of Virginia—undergraduate, graduate, and postdoctoral—who provided assistance, suggestions, and constructive criticisms: Brian Cadieux, Andy Kogelnik, Rakesh Nagarajan, Rick Gray, Jim Angstadt, Pat Mangan, and Craig Hocker. Thanks also to three other members of the Biology Department who played important roles in the development of *NeuroDynamix:* Tom Breeden, Dave Hudson, and Hal Noakes.

We thank the biology faculty and their many students who used and evaluated earlier versions of the *NeuroDynamix* manual and software. Although many colleagues have invested their time exploring this system, we wish to thank particularly Jim Angstadt and his students at Siena College; Henrik Jahnsen, Allan Djo, and their students at the University of Copenhagen; Ron Seaman and his students at Louisiana Tech University; Dennison Smith and his students at Oberlin College; Carl Thurman and his students at the University of Northern Iowa; Catherine Carr and Suzanne Dykstra at the University of Maryland; Ann Kammer at Arizona State University; and Richard Olivo at Smith College. We also convey our special thanks to the classes of graduate and undergraduate students in the Neurobiology course of the Biology Department at the University of Virginia who have cheerfully accepted the use of *NeuroDynamix,* during its development, as part of their instructional material during the past five years.

We thank Don Jackson, Executive Editor for Science and Medicine at Oxford University Press, for continual support; Janet Hronek for expert editorial work; Vicki Hoffmann for layout; and Lillian Hastie for packaging the software to fit with the written text. Editorial review of an early version of the text and program was kindly provided by Barbara Miller of Tempe, Arizona. The line drawings were generated by our talented illustrator, Boris Starosta.

The development of *NeuroDynamix* was materially assisted by funding from an NIH research grant (RO1-NS21778) and by a Fogarty Senior International Fellowship (FO6-TWO1924); the latter supported the Sesquicentennial Leave—which we gratefully acknowledge—for one of us (WOF) at Cambridge

University. We thank Peter Mathias, Master, and the Fellows of Downing College for selecting one of us (WOF) as the Thomas Jefferson Visiting Fellow during the spring of 1993. Their generous provision of a room with a view at the college helped inspire much of the final text. The NSF Center for Science and Technology (Center for Biological Timing; NSF grant DIR8920162) at the University of Virginia receives our appreciation for providing both office space and financial support.

We want to express our gratitude for the very able service of the staff and the active support of many colleagues in the Biology Department at the University of Virginia. We single out for particular mention our close colleagues Gene Block, Mike Mellon, and Masashi Kawasaki. Always available and of immeasurable aid was the loving support from our family and friends. Finally, we honor the love and encouragement of Lynette Friesen, wife and mother.

Section I

Neurophysiology and Modeling Exercises

Section I.1 Fundamentals of Electricity

I.1.1 INTRODUCTION

The pioneering work of electrophysiologists over the last century has led to the insight that the functions of neurons and of the nervous system can be usefully described by an analogy (the parallel-conductance model) with electrical circuits. This section provides an introduction to the electrical terms and concepts that provide the bases for understanding this analogy. We begin with a short description of the electrical components that make up circuits, namely, conductors, batteries, resistors, and capacitors. The basic quantities that describe these circuits are current, voltage, resistance, and capacitance. We then show how these terms are used in the descriptions of electrical circuits. Finally, we derive a few equations that are fundamental for understanding the parallel conductance model and its relationship to cell membranes.

I.1.2 ELECTRICAL DEVICES AND UNITS OF MEASURE

I.1.2.1 Current

Motion of electrons in wires (conductors) and ions in aqueous solutions (also conductors) is analogous to the motion of water in streams. The current in a stream (liters per second flowing by some point) has its analog, electrical current, in the flow of electrons or ions passing by some point (charges per second). The fundamental unit for electrical current is the Ampere (1.0 A is equal to the flow of 1 Coulomb/second). Because this current is very large where cells are considered, we use much smaller units, including milliamperes (1.0 mA = 10^{-3} A), microamps (1.0 μA = 10^{-6} A), nanoamps (1.0 nA = 10^{-9} A), and picoamps 1.0 pA (1 pA = 10^{-12} A). In equations describing electrical circuits, current is designated by the symbol I. Electrical devices that generate currents of controlled amplitude are represented by two interlocking rings (fig. I.1-1a).

Current is a directional (signed) quantity; the positive direction is that direction in which *positive* charges move (for current in a wire, the positive direction is the direction *opposite* to that in which electrons are moving). In aqueous solutions, currents are the result of movements by ions. Positive ions such as sodium (Na^+), potassium (K^+) and calcium (Ca^{2+}) move in the direction of positive current, whereas negative ions such as chloride (Cl^-) travel in the direction opposite to that of positive current. For currents through cell membranes the convention is to assign a negative value to currents flowing into the cell. For example, Na^+ ions often flow into cells; this is considered to generate a negative current. The *outward* flow of K^+ through the cell membrane, a frequently encountered current, is assigned a positive value. Finally, the *inward* flow of Cl^- ions also generates a positive current (even though chloride ions are negatively charged). As a taste of things to come, you may note that inward (negative) currents have an excitatory effect on neurons, whereas outward (positive) currents are inhibitory.

(a)

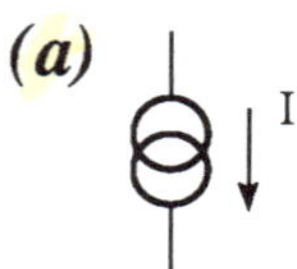

(b)

E

(c)

(d)

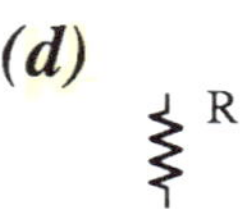

(e)

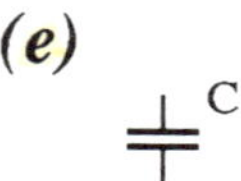

Figure I.1-1 Symbols for electrical components. *a*) Source of constant current. The arrow indicates the direction of the current generated by the source. *b*) Battery. This is a source of constant voltage. The "+" sign indicates the positive side of the battery; the "E" is used to indicate the battery potential. *c*) Circuit ground. Circuit lines directly connected to this symbol have zero electrical potential. *d*) Resistor. Current through the resistor generates a voltage given by Ohm's law. The value of *R* designates the resistance value. When the value of the resistance (conductance) is not fixed, a diagonal arrow is drawn through the symbol. *e*) Capacitor.

I.1.2.2 Voltage

To continue the hydrological analogy, we can define the term "voltage" as a measure similar to water pressure, namely, electrical pressure. Just as falling water is capable of work (turning a water wheel), so electrical pressure (electrical potential or voltage) provides the capability of electrical work. To be strictly precise, voltage refers to the potential difference between two points. However in common usage this somewhat awkward compound term is simply reduced to "potential." The fundamental unit of electrical potential is the Volt (V). This unit again is too large for most measurements in bioelectricity; hence, we commonly employ smaller units, such as the millivolt ($1.0 \text{ mV} = 10^{-3} \text{ V}$) and the microvolt ($1.0\ \mu\text{V} = 10^{-6} \text{ V}$). It is sometimes a source of confusion for students that the symbol for voltage in equations, *V*, is also the symbol for the unit of measure for this quantity. Note that to designate the variable name we use an italicized "*V*" and that to designate the unit of measure we simply use a capital "V." Batteries, which are electrical devices that generate constant voltages, are represented in circuits by two parallel lines of unequal length, with the longer line representing the positive (+) terminal (fig. I.1-1b). The symbol for the potential difference across a battery is *E*.

Like current, voltage (or potential) is a signed quantity. Because voltage is the difference between two points, we should actually write V_{ab} (*V* is the variable here) to show explicitly that we are describing the difference in the potential between points "a" and "b"; i.e., $V_{ab} = V_a - V_b$. With this explicit terminology, value V_{ab} is positive if the potential at "a" is greater than at "b." (That is, point "a" is upstream from point "b".) In a resistor (an electrical circuit component that allows electron or ion flow), the direction of the current (like the water in our stream) is always from the higher to the lower potential; in other words, current in a resistor always flows from the end with the more positive potential to the end with the less positive potential. To assist in making the comparison between local potentials required for determining voltage, we have a special symbol to represent a point in an electrical circuit with 0 potential (electrical ground)—three parallel lines, each shorter than the last (fig. I.1-1c).

An important measure for electrophysiologists is the potential difference (V_m) between the inside and the outside of cells, where $V_m = V_{inside} - V_{outside}$. It is common usage to set $V_{outside}$ to zero. Because only the *difference* can be measured in any case, this convention does not alter the size of the measured potential. When the parallel conductance model is used to represent electrophysiological function, the ground symbol is often included in order to show explicitly that the outside of the cell is assumed to be at zero, or ground potential. Because of this usage, V_m is simply equal to V_{inside}.

I.1.2.3 Resistance

Resistance is a measure of the hindrance to charge movement presented by an electrical component called a "resistor." (We shall describe the properties of a resistor in more detail below.) The fundamental unit of resistance is the ohm (Ω). Water flowing through pipes or faucets encounter resistances that are analogous to the electrical resistance; smaller pipes offer a higher resistance. For biological measurements the ohm is a very small unit of measure. More appropriate units are kilohms (1.0 KΩ = 10^3 Ω); megaohms (or megohms; 1.0 MΩ = 10^6 Ω) and even gigaohms (1.0 GΩ = 10^9 Ω). Resistance is a *scaler* quantity; it has no sign or direction (i.e., it is always positive).

Electrophysiologists often describe resistors by using the term "conductance" to designate the ease with which ions pass through membranes or move in the cell cytoplasm rather than resistance. Conductance is the reciprocal of resistance; physical resistors that have a high resistance have a low conductance. (In our plumbing analogy, small pipes have a high resistance to water flow and therefore a low conductance.) In symbols, $g = 1/R$, where g is the symbol for conductance. It is worth noting that the resistance of resistors connected in series add linearly. Similarly, the conductances of resistors arranged in a parallel relationship add. The fundamental unit of conductance is the siemen (S). Smaller units are the millisiemen (1.0 mS = 10^{-3} S), the microsiemen (1.0 μS = 10^{-6} S), the nanosiemen (1.0 nS = 10^{-9} S), and the picosiemen (1.0 pS = 10^{-12} S). In equations describing the current through a resistor or the potential difference across a resistor (between the two ends), the value of the resistance is designated by the symbol R. Physical resistors, which have a fixed value of resistance to current flow, are represented in circuits by a zigzag line (fig. I.1-1d). Devices whose resistance can be altered (potentiometers) are designated by a resistor symbol with a superimposed diagonal arrow.

I.1.2.4 Capacitance

A capacitor is a device for storing electrical charge. Physically, capacitors often consist of two conductors separated by a thin insulating film (a dielectric) that has very high electrical resistance. (In our water analogy, the capacitor may be thought of as an enlarged section of garden hose whose lumen is blocked in the center by a flexible diaphragm.) The value of a capacitor is directly proportional to the area of the two conductors and inversely proportional to the thickness of the insulator. Because the lipid bilayer of cell membranes is a very thin insulator separating two conductors, the extracellular and intracellular fluids, the cell membrane is an excellent capacitor.

The fundamental unit of capacitance is the farad (F), a very large unit of electrical storage capacity. Smaller, commonly encountered units are the microfarad (1.0 μF = 10^{-6} F), the nanofarad (1.0 nF = 10^{-9} F), and the picofarad (1.0 pF = 10^{-12} F). The symbol for a capacitor is two parallel lines of equal length (fig. I.1-1e); in equations the value of a capacitor is designated by the letter C.

I.1.3 RELATIONSHIPS BETWEEN CURRENT AND VOLTAGE

Electrophysiologists can make two types of direct measurements. The ammeter (symbolized by a circle enclosing the letter "A") measures both the magnitude and the sign of electrical currents. The voltmeter (circle enclosing a "V") detects electrical pressure, again measuring both magnitude and sign. These are assumed to be perfect devices in that the ammeter is assumed to be a perfect conductor (with zero resistance) and the voltmeter is assumed to be a perfect resistor (with zero conductance).

I.1.3.1 Ohm's law

To understand the basics of electric currents and the resulting potential differences across resistors, we begin with Ohm's law. This law states that the potential difference across a resistor is proportional to the current that passes through the resistor. Consider the experimental setup shown in figure I.1-2a, which depicts a current source, a resistor, an ammeter (A), and a voltmeter (V). The two ends of the resistor are labelled by "a" and "b," with the ground (0 voltage) at point b. The direction of the current through the resistor is from "a" to "b" (indicated by the arrow near the current source). Our voltmeter will indicate that the voltage V_{ab} is positive; that is, point "a" is at a higher potential than at point "b".

(a)

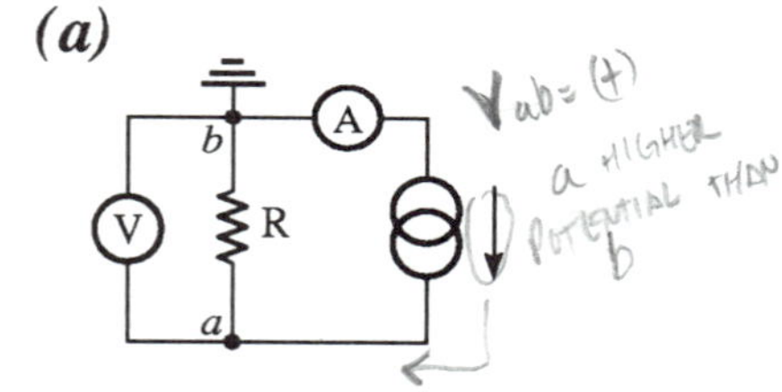

(b)

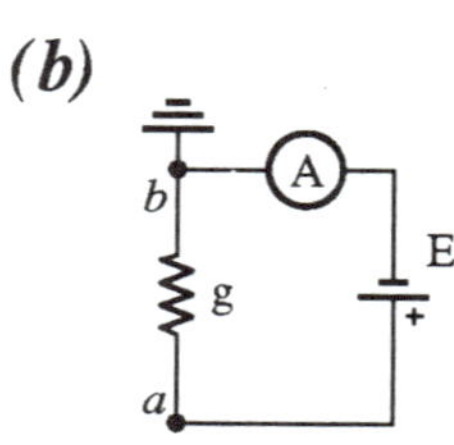

Figure I.1-2. Circuits for verifying Ohm's law. *a*) Circuit for finding the value of a resistor by applying a current. *b*) Circuit for finding the conductance of a resistor by applying a voltage. The potential at point "b" in each circuit is zero because this node is connected to ground.

Written in symbols (mathematically), Ohm's law states that

$$V_{ab} = R * I, \tag{I.1-1}$$

with R a constant that describes the proportionality between the applied current (I) and resulting voltage (V_{ab}) across the resistor. Ohm's law as stated in equation (I.1-1) is actually an empirical result obtained by graphing V_{ab} against I. In this graph R is the slope of a straight line (see modeling exercise Elect-1). Note that with point "b" set to zero, the voltmeter simply gives the value at point "a," and V_{ab} is written simply as V. Hence, we can write Ohm's law in shortened notation as

$$V = R * I. \tag{I.1-2}$$

For electrophysiologists the reciprocal of resistance, conductance, is often a more convenient description of neuronal function. As noted above, the relationship is $g = 1/R$. Using this relationship we can rewrite Ohm's law as

$$I = g * V. \tag{I.1-3}$$

This new equation implies that a graph of current against the voltage applied to a resistor should yield a straight line whose slope is g. Figure I.1-2b illustrates the experimental setup to determine the value of the conductance of a resistor. Ohm's law as embodied in equation (I.1-3) provides the basis for our description of the electrical properties of cell membranes (see modeling exercise Elect-2).

I.1.3.2 Kirchhoff's rules for electrical circuits

The electrical circuits that describe bioelectricity usually include several current paths, raising the problem of determining the size of currents in a path. Kirchhoff's two rules were formulated to solve this problem. These rules in essence state 1) that charges are neither created nor destroyed in electrical circuits (charge conservation) and 2) that when traversing a loop in an electrical circuit, the beginning and ending points have the same electrical potential.

The first current-node rule applies when a circuit has a node (also called a "branch point") at which three or more conductors meet. It states that *the algebraic sum of the currents toward any node is equal to 0.* That is, in shorter notation, $\Sigma\ I_j = 0$ at any node. This rule is applied by first assigning a preliminary direction to all currents in the conductors that meet at the node. The sum of the currents is then formed by assigning positive values to currents directed into the node and negative values to currents directed away from the node. For example, as illustrated in figure I.1-3, the currents flowing into the node "a" must equal those flowing out; in other words

$$I_1 - I_2 - I_3 = 0 \tag{I.1-4}$$

Note that in this example, currents 2 and 3 sum as negative quantities because the arrows for these currents are directed away from the node, whereas I_1 sums as a positive quantity because the current arrow is directed toward the node. For complex circuits (including those encountered in neuronal models) several independent node equations may be obtained from a circuit. In general, the number of independent current node equations is one less than the number of nodes.

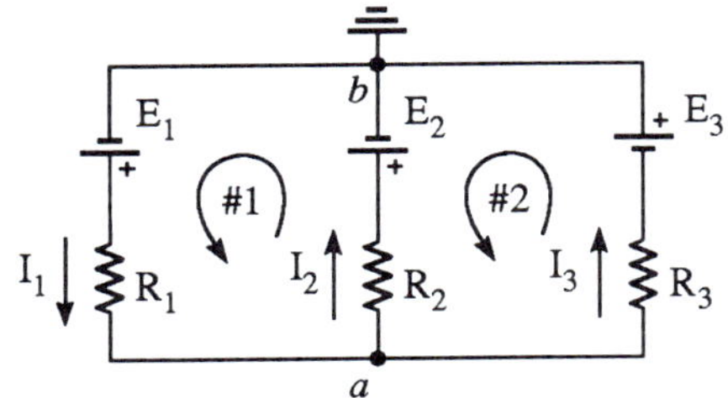

Figure I.1-3. Parallel resistors. The three resistors (R_1, R_2, and R_3) each in series with a battery are connected to form a parallel circuit. Straight arrows show the (assumed) direction of current through each of the resistors. Circuit loops #1 and #2 are indicated by the curved arrows.

Kirchhoff's loop rule states that *in any closed loop in an electrical circuit the sum of the potential drops in the resistors of the loop and the potential (voltage) gains due to batteries is equal to zero.* That is, in going around a closed loop in an electrical circuit, $\Sigma(R_j * I_j) + \Sigma E_i = 0$, where the term "$R_j * I_j$" is the potential difference across the jth resistor with resistance R_j due to a current I_j in that resistor and where E_i is the value of the ith battery. This second rule embodies the principle that in traversing a closed loop the sum of the voltage drops is equal to the sum of the voltage gains. (The same principle applies to elevation gains and losses when completing a *circuit* hike in the Shenandoah National Park.)

The loop rule is applied as follows (fig. I.1-3). First, assign a preliminary direction to the current in each loop in the circuit (use the directions chosen for the node rule). Second, choose a direction for going around each loop. Third, mentally travel around each loop. Whenever you encounter a resistor, use a *positive* value for the voltage drop ($+R_j * I_j$) if you are traveling in the direction of the current. When you encounter a battery, use a *negative* sign ($-E_i$) if you travel from the negative to the positive side of the battery (climbing uphill is a negative experience!) and a *positive* sign ($+E_i$) if you travel from a positive to a negative side. The sum of the potential drops across the resistors added to the sum of the (signed) battery potentials is then set to equal zero. So for loop 1, beginning at node "a" (fig. I.1-3),

$$R_2 * I_2 + E_2 - E_1 + R_1 * I_1 = 0. \quad \text{(I.1-5)}$$

The equation for loop 2, again beginning at node "a," is similar, namely,

$$R_3 * I_3 - E_3 - E_2 - R_2 * I_2 = 0. \quad \text{(I.1-6)}$$

The single application of the node rule and the double application of the loop rule provide us with three equations in three unknowns for the circuit shown. If the values of the batteries and of the three resistors are given, we can use these three equations to solve for the three unknown currents. If the values of the batteries and the currents are given, we can solve for the values of the resistors. The solutions to the equations describing the circuit may result in currents with negative signs. A negative sign for a current indicates that the direction of that current is in the direction opposite to the original (arbitrary) assignment (see modeling exercise Elect-3).

I.1.4 ROLE OF MEMBRANE CAPACITANCE

The excellent capacitor formed by the cell membrane and the surrounding fluids has a profound effect on the electrical potentials observed in cells. In the absence of a capacitor, potentials across resistors change instantly when a current is applied (see modeling exercise Elect-1). The effect of the membrane capacitor is to slow the rate at which potentials change, as we shall now see.

As described above, capacitors are charge storage devices that are charged up and discharged by electrical currents. The charge (Q—measured in Coulombs) on a capacitor is directly proportional to the voltage applied. In symbols,

$$Q = C * V, \quad \text{(I.1-7)}$$

where the proportionality constant C is defined as the value of the capacitance and V is the applied voltage. We can find the value of current flowing into a capacitor (I_C—the "displacement current") by observing that current is just the rate of charge movement, i.e., that $I_C = \mathrm{d}Q/\mathrm{d}t$, the derivative of the charge with respect to the time t. Hence, taking the derivative of both sides of equation (I.1-7), we have

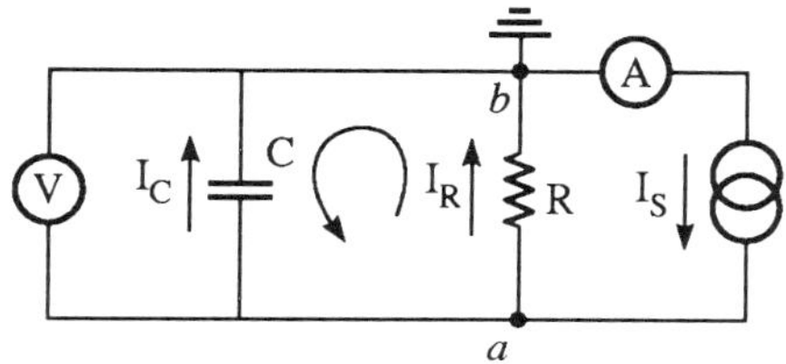

Figure I.1-4. Simple *R-C* Circuit. The circuit consists of a single loop formed by a capacitor *C* and resistor *R* wired in parallel. Current (I_S) is supplied by the current source; its amplitude is measured by the ammeter (A). The voltage across the circuit is measured by the voltmeter (V).

$$I_C = \mathrm{d}Q/\mathrm{d}t = C * \mathrm{d}V/\mathrm{d}t. \qquad (I.1\text{-}8)$$

When determining the values of currents in electrical circuits with a capacitor, we can use the right side of equation (I.1-8) to describe the current through the capacitor. The value of the displacement current I_C is positive if the current acts to induce a more positive voltage across the capacitor. In other words, the side of the capacitor into which current is flowing becomes more positive with time.

We will now apply Kirchhoff's rules and equation (I.1-8) to find the values of currents and potentials in a circuit consisting of a current source, a resistor, and a capacitor (fig. I.1-4). This simple circuit represents a piece of cell membrane with equal concentrations of ions on the two sides. There are two current nodes in this circuit, "a" and "b"; hence, we can use the node rule once to give us at the "a" node

$$I_S - I_R - I_C = 0, \qquad (I.1\text{-}9)$$

where I_S is the value of current from a constant-current source, I_R is the value of the current through the resistor, and I_C is the displacement current through the capacitor.

We can apply the loop rule once beginning at "a" and following the arrow around the loop to generate the equation

$$I_R * R - V_C = 0, \qquad (I.1\text{-}10)$$

where V_C is the voltage across the capacitor. Note that V_C is equal to V, the voltage between points "a" and "b" in the circuit. Also, from Ohm's law, the current across the resistor (I_R) is equal to V/R. Rearranging equation (I.1-9) and combining it with equation (I.1-10) we have

$$I_C = I_S - I_R = I_S - V/R. \qquad (I.1\text{-}11)$$

The term on left side of this equation is described by equation (I.1-8); hence, we may write

$$C * \mathrm{d}V/\mathrm{d}t = I_S - V/R. \qquad (I.1\text{-}12)$$

Now multiply both sides of this equation by R. We now define a new symbol "τ" (tau), which is the product of the resistance and capacitance ($R * C$). The term τ is called the "time constant" of the circuit and describes how quickly membrane potential changes when a current is applied. We also define the term V_0 as $I_S * R$, where V_0 is either the initial or the steady-state value of the potential across the circuit. With these new symbols equation (I.1-12) becomes

$$\tau * \mathrm{d}V/\mathrm{d}t = V_0 - V. \qquad (I.1\text{-}13)$$

This linear, first-order differential equation has two solutions, depending on the initial conditions. For the initial condition in which the capacitor is uncharged and the current I_S is turned on at time $t = 0$, we have the system equation

$$V = V_0 * (1 - e^{-t/\tau}). \qquad \text{(I.1-14)}$$

For this initial condition, the voltage between points "a" and "b" (fig. I.1-4) described by this equation is a rising curve with a time constant τ equal to $R * C$ and with a steady-state value of V_0 (= $I_S * R$). On the other hand, if we have an initial steady-state condition with the capacitor charged and *then* (at time $t = t_0$) turn off the current, we have the system equation

$$V = V_0 * e^{-t/\tau}. \qquad \text{(I.1-15)}$$

This second equation describes V as a falling exponential with an initial value of V_0 (= $I_S * R$) and with a steady-state value of zero. The time constant τ describing the rate of change in the potential across the circuit again is $R * C$.

Equations (I.1-14) and (I.1-15) inform us that with the addition of a capacitor in parallel to the resistor, the voltage across the resistor does not assume a new final value instantly. Instead, the potential changes gradually, with the rate of change set by the values of the resistor and the capacitor. That is, the effect of the capacitor in the circuit of figure I.1-4 is to slow the rate at which voltage can be changed by the current source.

Please note that when carrying out calculations *always first convert quantities to their fundamental units*. Then do the calculations. Finally, reconvert to the units appropriate for neurophysiological measurements (see modeling exercises Elect-4 and Elect-5).

Modeling Exercises

NeuroDynamix Model: RESISTOR Exercises: Elect-1 to Elect-5

For all modeling exercises, load the indicated MODEL and then use SETUP to configure the model for the designated modeling exercise. For some of the exercises, when proceeding from one exercise to the next, it is advisable to stop the graphing at the right-hand edge of the graph window before opening the SETUP window. If the windows should become excessively cluttered and confusing, simply type "q" or click the mouse on the DOS icon to exit the program; then type "ND" to restart NeuroDynamix.

Introduction

This set of exercises is designed to demonstrate the fundamentals of electricity as they relate to neurobiology. The experiments introduce the components of electrical circuits—conductors, resistors, batteries, and capacitors. For these exercises the components are assembled into a variety of simple electrical circuits. Note that because these exercises are meant to simulate neuronal properties, resistors are described in terms both of their resistance and of their conductance (the reciprocal of resistance values).

Elect-1 Ohm's law (resistance)

For this exercise RESISTOR is configured to consist of a single resistor connected to a current source (stimulator) with a voltmeter (oscilloscope) connected to the ends of the resistor to monitor the voltage generated by the current through the resistor (fig. I.1-2). When you begin this exercise the computer screen displays four windows: 1) a graph of voltage (*Vtot*) versus current (*Stimulus*) on the PHASEPLANE window; 2) the MINIMIZED Stimulator window; 3) the PARAMETER window; and 4) a TIMESERIES window (titled "Elect-1") showing the instantaneous voltage (*Vtot*) across the resistor and the current (*Stim*) through the resistor (fig. I.1-5).

The purpose of this exercise is to demonstrate Ohm's law that $V = R * I$, where V is the voltage across a resistor, R is the value of the resistance, and I is the applied current. Ohm's law implies that when current is passed through a resistor, a voltage is generated that is proportional to the imposed current. Therefore, if a graph is generated in which the abscissa (x-axis) is the current and the ordinate is the voltage, the graph should be a straight line with slope R.

In this exercise the stimulator is initially set to "On" and the current amplitude is 0. Click on the STOP/GO icon to begin the exercise and change the current amplitude from 0 to 10 nA in 1 nA steps. After each step observe the values of current and voltage on the TIMESERIES graph on the right and the graph of V vs I at the left (PHASEPLANE window). Calculate the slope of the line generated in the PHASEPLANE window to verify that it does equal the value of R given in the PARAMETER window. (Remember that current [I] is given in units of nA [10^{-9} A] and that the voltage [*Vm*] is in units of millivolts

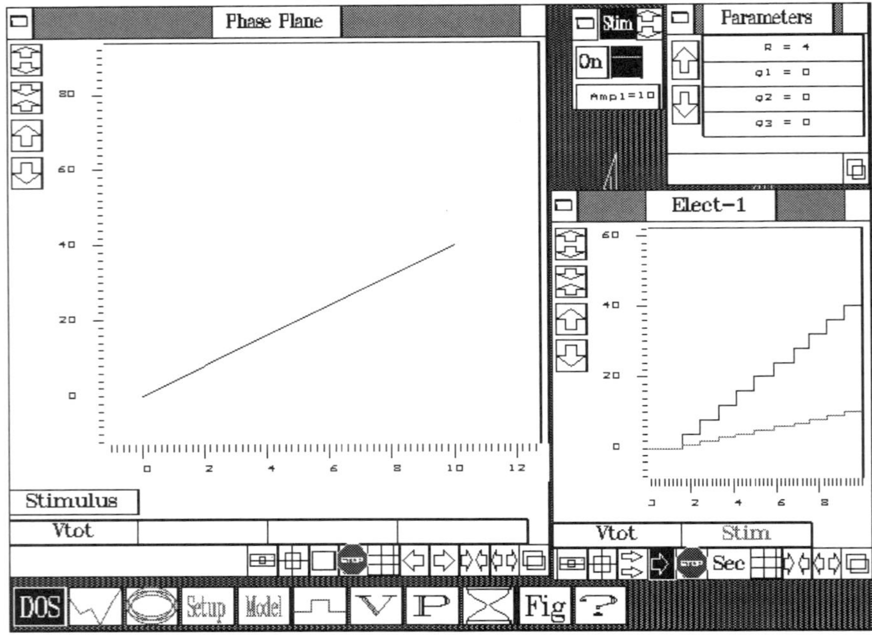

Figure I.1-5. Monitor display for exercise Elect-1

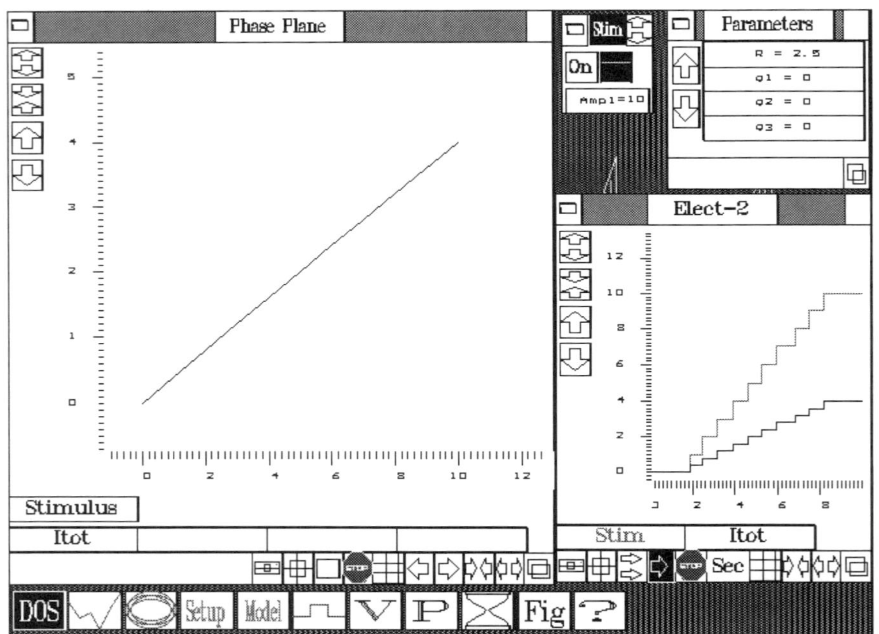

Figure I.1-6. Monitor display for exercise Elect-2

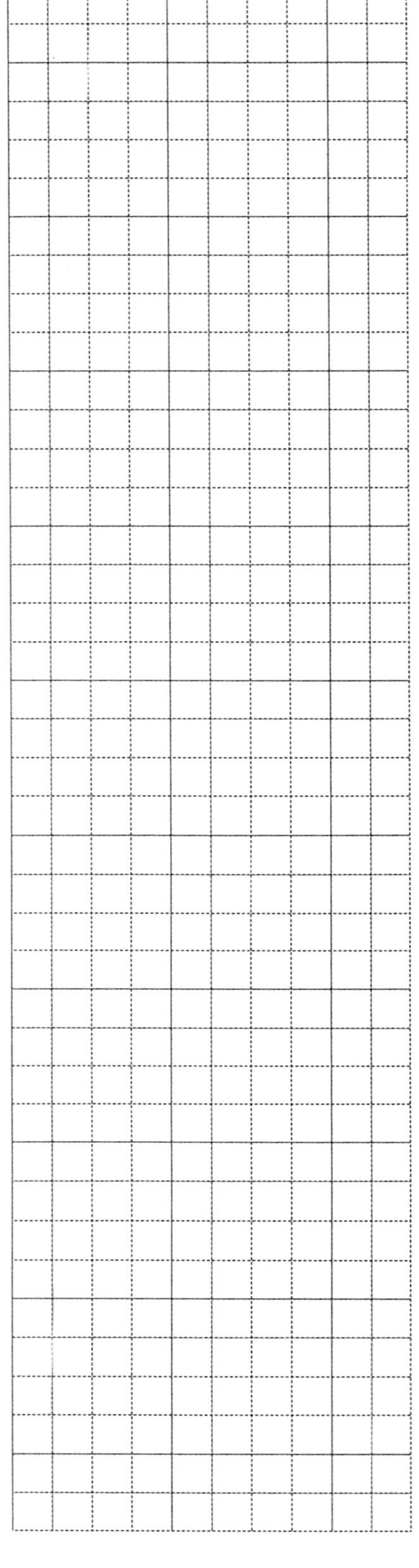

[10^{-3} V]; hence, the resistance value [R] is in units of Megohms [10^6 Ω].) The I-V graph can be generated automatically by setting the stimulator waveform to a triangular shape (click on the stimulator waveform icon until you see a triangular waveform. Using this stimulator waveform and a current amplitude of 10, make I-V graphs for both larger and smaller values of the resistance (try 2 MΩ and 8 MΩ). Notice that the new graphs have slopes that are double and half of those in the original graph. Also notice that in this very simple, ideal (and unrealistic) circuit voltage changes across the resistor occur instantly as the current amplitude is altered. We will see later that when capacitance is present in the circuit changes in voltage are slower.

Elect-2 Ohm's law (conductance)

For this exercise RESISTOR again is configured to simulate a single resistor. Now, however, the resistor is connected to a *voltage* source (the stimulator now supplies a predetermined voltage), and an ammeter is connected to measure the current through the resistor (fig. I.1-2b). The same windows are open as in Elect-1, but note that the axes are different (fig. I.1-6).

This exercise provides a second demonstration of Ohm's law; namely, $I = V/R$. In other words, the current through a resistor is proportional to the electrical pressure (voltage) applied. Hence, a graph with current on the ordinate and voltage on the abscissa should be a straight line with slope $1/R$. For just this situation, we introduce the term "conductance," defined as $g = 1/R$. Thus, the slope of the V-I graph is simply g.

Begin the exercise (after clicking on the STOP/GO icon) by increasing the voltage (*Ampl* on the stimulator) from 0 to 10 mV in 1 mV steps. After each step note the values of voltage and current on the TIMESERIES graph at the right and the graph of I vs V at the left (PHASEPLANE window). Calculate the slope of the line generated in the PHASEPLANE window to verify that it does equal the reciprocal of R given in the PARAMETER window.

(a)

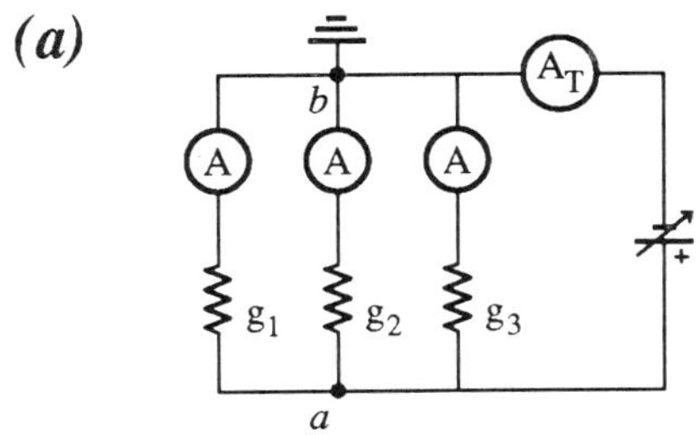

Figure I.1-7a. Parallel conductance circuit. The three conductances (g_1, g_2, and g_3) are connected in parallel to the variable voltage source (indicated by the battery with the diagonal arrow). The three ammeters measure the currents through each of the conductors and the total current (A_T) flowing in the circuit.

Elect-3 Parallel conductances

For this exercise RESISTOR is configured to simulate three conductances (that is, three resistors) connected in parallel to a common voltage source. Ammeters are connected to measure the currents (*Ig1*, *Ig2*, and *Ig3*, in units of μS) through each conductance, and another ammeter (*Itot*, in nA) is connected to measure the total current through the three resistors. A source of potential, a "variable battery," is connected to generate a series of voltages across the resistors (fig. I.1-7a). When you begin this exercise, the computer screen again displays four windows: 1) a PHASEPLANE window for graphing currents *Ig1*, *Ig2*, *Ig3*, and *Itot* versus voltage (*Stimulus*); 2) the MINIMIZED stimulator window; 3) the PARAMETER window, giving the values of the three conductances (*g1*, *g2*, and *g3*); and 4) the TIMESERIES window, showing the instantaneous voltage (*Stim*) and the total current (*Itot*) passing through all three conductances (fig. I.1-7b).

(b)

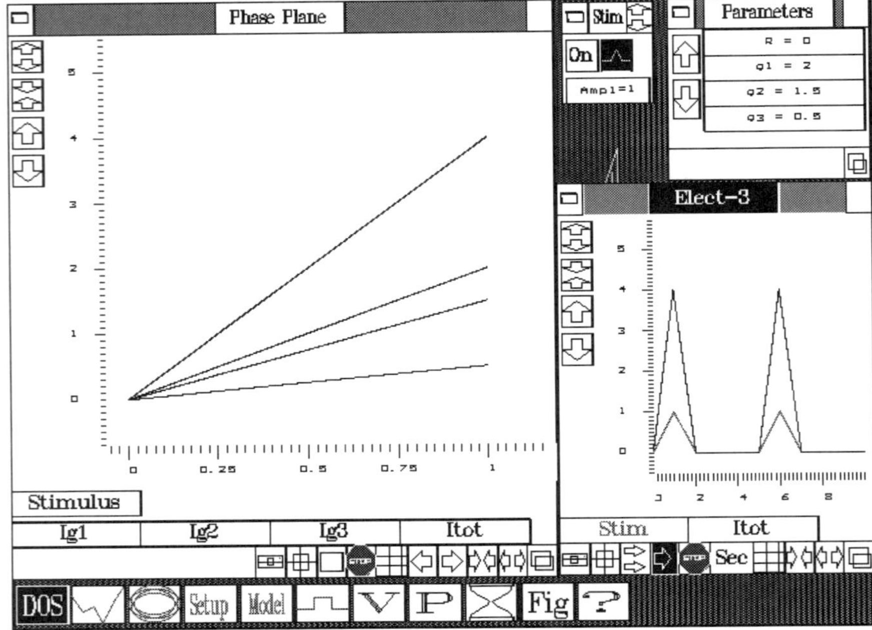

Figure I.1-7b. Monitor display for exercise Elect-3

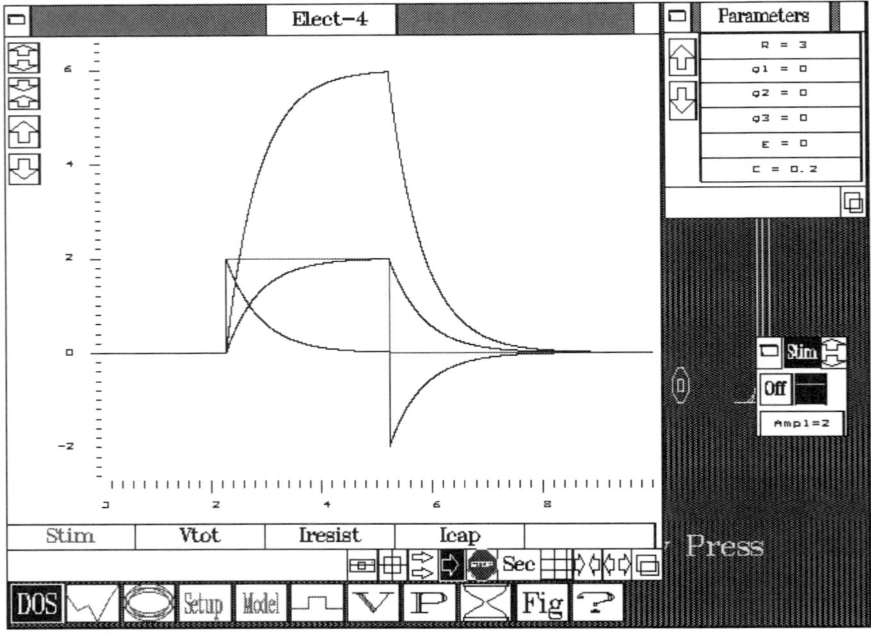

Figure I.1-8. Monitor display for exercise Elect-4

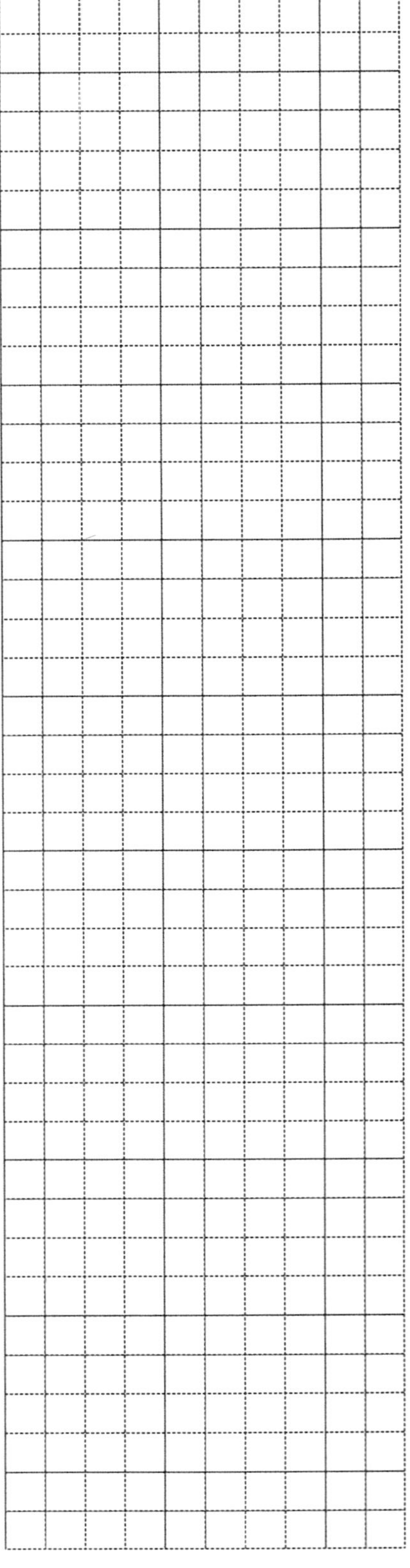

The purpose of this exercise is to demonstrate that when resistors are wired in parallel, it is useful to use conductance, rather than resistance, as a measure of the ease with which current passes through the circuit. Remember that for resistors in parallel the circuit resistance is equal to the reciprocal of the sum of the reciprocals but that conductances simply add; i.e., $gtot = g1 + g2 + g3$. Verify this relationship by finding *gtot* (equal to the slope of the *Itot* graph on the phase plane) and showing that this value equals the sum of the three conductances (find the slope of each of three individual current traces on the phase plane.)

To get a better feel for how conductance values determine current flow, vary the values of *g1*, *g2*, and *g3* individually. Does increasing the value of the conductance increase or decrease the current through the resistor (conductance)? How is the total current affected? Notice again that the amplitudes of all currents change instantaneously when the applied voltage is altered.

Elect-4 Current-voltage relationships in *R-C* circuits

For this exercise RESISTOR is configured to consist of a single resistor connected in parallel to a capacitor. A current source (the stimulator) is set up to generate a 2 s current pulse when the stimulator is turned on (fig. I.1-4). The large TIMESERIES graph shows the instantaneous values of the current applied (*Stim*, in units of nA), the voltage across the circuit (*Vtot*, in mV), the current through the resistor (*Iresist*, in nA), and the current through the capacitor (*Icap*, in nA) (fig. I.1-8).

The purpose of this exercise is to demonstrate that the presence of a capacitor in electrical circuits (and in neurons) slows the rate at which voltages (and neuronal membrane potentials) change when the applied current changes. This exercise also demonstrates that we can describe this change in rate by a single constant, τ (tau) $= R * C$, the product of the resistance and the capacitance values.

Begin the exercise by clicking the mouse on the STOP/GO icon and letting the trace move for 1 s of simulated time. Then click the mouse on the ON/OFF icon on the stimulator to pass 2 nA current through the circuit. Turn the stimulator off after 2 s of model time and wait until the trace stops at the right edge of the graph. Look at the graph carefully. First, you should observe that the voltage across the resistor does not change instantly; rather, it rises exponentially to a steady-state value. The reason for this gradual change in voltage is that the current through the resistor increases gradually (exponentially) after the current is turned on. To understand these curves observe that when the stimulator is turned on all stimulus current passes through the capacitor ($Icap = Stim$), whereas no current passes through the resistor ($Iresist = 0$). With time, the current through the capacitor decreases while the current through the resistor increases; the sum of these two currents is constant, always equal to the stimulator current. Eventually, the current through the capacitor is zero and the current through the resistor is equal to the stimulator current. When this steady state is achieved, the voltage across the resistor (*Vtot*) becomes constant; it equals $Stim * R$. Also study the currents and voltage when

(a)

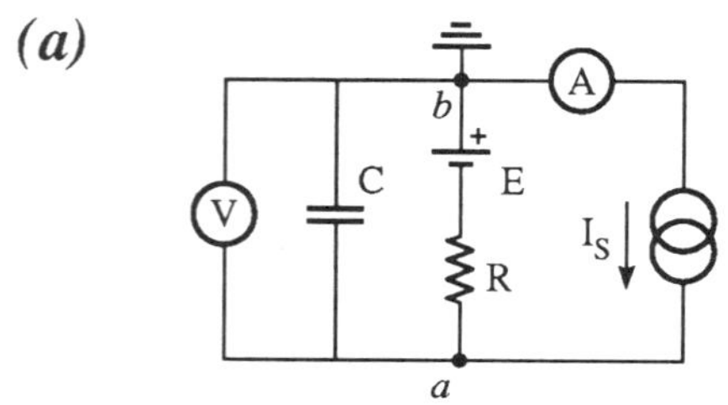

(b)

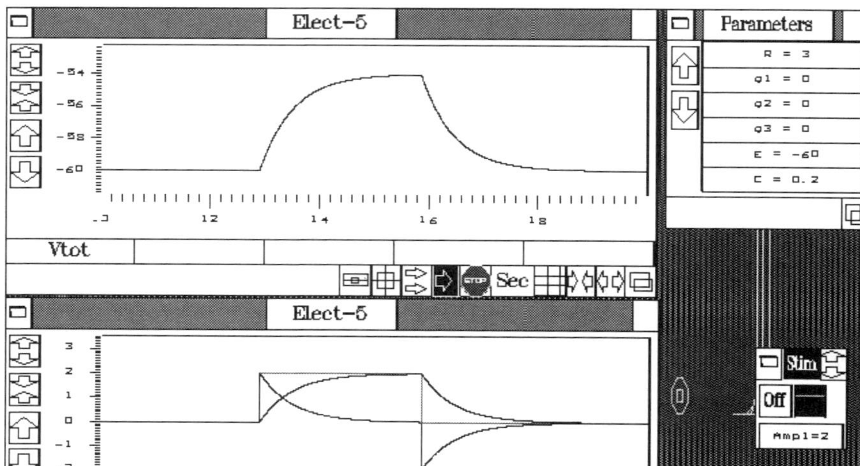

Figure I.1-9. Membrane analog for a simple neuron. *a*) Circuit diagram. *b*) Monitor display for exercise Elect-5.

the stimulator is turned off—remember that the sum of the currents through the resistor and capacitor are equal to the stimulator current (i.e., their sum is zero when the stimulator is off).

The time constant of an *R-C* circuit describes the rate at which membrane potential changes when a current is applied or turned off. Recall, by referring to equation I.1-15, that if we take time t as zero when the stimulator is turned off, then at $t = \tau$ the voltage $V = V_0 * e^{-1}$; or $V = V_0 * 0.37$. To find the time constant τ open the ANALYZE window. Click the mouse twice at the beginning of the voltage drop (just where you turned the stimulator off) and set the cursor cross-hairs to that part of the *Vtot* curve at which it equals 0.37 times the steady-state value. Click the mouse again; then read the value of τ from *Delta X* value in the ANALYZE window. Confirm your experimental result by finding the theoretical circuit time constant by multiplying $R * C$ shown in the PARAMETER window. The values of capacitance in this simulation are express in μF. Learn more about the role of the membrane capacitance in slowing the rate of voltage changes in electrical circuits by repeating these exercises with both larger and smaller values of the capacitance C.

Elect-5 Equivalent circuit for cell membranes

For this exercise RESISTOR is configured to consist of a single resistor connected in series with a battery; these two elements are then in parallel with a capacitor. A current source (the stimulator) supplies external current to the circuit while a voltmeter measures the voltage across the circuit (fig. I.1-9a). Four windows are displayed: 1) an upper TIMESERIES window that graphs the voltage across the circuit (*Vtot*); 2) a lower TIMESERIES window showing current applied (*Stim*), current through the resistor (*Iresist*), and current through the capacitor (*Icap*); 3) the STIMULATOR window; and 4) the PARAMETER window (fig. I.1-9b).

The purpose of this exercise is to illustrate the link between electrical circuits and their usefulness in representing the transmembrane potentials in neurons. You will see that the only consequence of adding a battery to the circuit employed in exercise Elect-4 is that there now is a resting potential (voltage) that is equal to the battery voltage. This battery voltage generates on offset to which all other potentials are added.

This exercise is a repetition of the first part of exercise Elect-4. Begin by clicking on the STOP/GO icon and letting the traces run all the way across the windows (to let transients decay). Click the mouse on the STOP/GO icon, letting the traces go about 1/3 of the way across the windows (until all variables are at their steady-state levels). Then turn the stimulator on to pass a positive current through the circuit. Watch how the voltage changes with time as the result of this step change in current. Note that the voltage across the circuits becomes less negative, going upward from its resting value. Now turn the stimulator off and observe the waveforms. Notice that the potential returns to its resting level. These waveforms again are exponential, following the exponential increase and decline of the current through the resistor. Calculate the voltage across the resistor and show that at all times the voltage across the combined resistor and battery is simply the algebraic sum of the battery voltage and the voltage across the resistor.

Section I.2 Patch-clamp Recording

I.2.1 INTRODUCTION

An exciting and fairly recent technical development in neurobiology is the patch-clamp recording technique. Patch-clamp recording was developed by Bert Sakmann and Erwin Neher during the 1970s to permit scientists to examine the functions of individual protein molecules that form ion channels in cell membranes. Sakmann and Neher received the Nobel Prize in 1991 for this outstanding contribution, which has revolutionized our thinking about the electrical properties of cell membranes. This section provides a brief introduction into our current thinking about how proteins form ion channels and describes the patch-clamp technique.

I.2.2 THE CELL MEMBRANE

Ever since the work of Schwann, and later that of Cajal, established that neurons are individual cells, the description of the boundary that separates the inside of a cell from the outside and from its neighbors has been an important research focus. That this boundary, the cell membrane in animal cells, is of critical importance becomes immediately obvious when we remember that the internal milieu of cells differs greatly from the extracellular environment. Not only are such large organelles as mitochondria and the endoplasmic reticulum retained within cells by the membrane, but even small polar molecules and small ions are retained or are selectively excluded. It is worthwhile to remember that the ionic concentrations differ greatly between the inside and the outside of cells. Although the absolute ionic concentrations for differing animal groups vary considerably, the concentration of potassium within cells is always much greater than in the surrounding fluids, and sodium ions are much less concentrated within cells than on the outside. Similarly, chloride and calcium ions are at a much higher concentration on the outside than on the inside of cells. These concentration imbalances form the bases for electrical signaling in neurons.

Cell membranes are composed of three principle components: lipids, proteins, and carbohydrates. (Although of critical importance for cell recognition and other cellular functions, we will not deal with carbohydrates in this book.) Recall that lipids are amphipathic molecules; they have long hydrophobic hydrocarbon tails that are joined to hydrophilic head groups composed of polar regions. As their contribution to the cell boundary, membrane lipids form a fantastically thin molecular bilayer film two molecules in width. This bilayer is only approximately 4 nm across. That's 4 millionths of a millimeter or roughly 4/100,000 of the thickness of this page! Lipid bilayers are self-assembling structures in aqueous solutions because, arranged as a bilayer, the hydrophobic lipids tails are associated with similar hydrophobic regions of other lipid molecules, while the hydrophilic polar heads are adjacent to other polar head groups or to polar water molecules. It should be clear that the lipid bilayer, with its hydrophobic center, provides a barrier impenetrable to the flow of ions; in other words, the lipid bilayer is a near-perfect resistor. Electrophysiology, however, is all about the flow of ions through membranes, so there must be membrane components that modify the resistive (or, to use the inverse term, the conductance) properties of the lipid bilayer. These modifying components are proteins found inside and outside of all cell membranes.

Integral proteins extend through the lipid bilayer of the cell membrane. These molecules have amphipathic structures, like the lipids, with the hydrophobic portions confined within the midregion of the lipid bilayer and with hydrophilic regions exposed both to the extracellular fluid and to the internal, cytosolic milieu. The integral proteins form a communication link between the insides and outsides of cells. In particular, the proteins provide a great variety of specific molecular routes by which ions pass through the otherwise impenetrable lipid barrier. One such route is the ion pumps, which utilize ATP as an energy source to convey ions—particularly sodium, potassium, and calcium—against concentration gradients to establish the ionic imbalance between the insides and outsides of cells. The action of ion pumps in setting up concentration differences on the two sides of a membrane is analogous to pumping water into an overhead tank or to charging a battery.

I.2.3 ION CHANNELS

As we discussed in section I.1, conductors (or resistors) are paths by which ions can flow from a higher to a lower electrical potential. Such paths, or ion channels, are formed by a variety of integral proteins found in cell membranes. Channel proteins are large molecules (about 240 kD for a complete, functional molecule) that are composed either of multiple subunits or a single, very long amino acid chain that includes multiple domains. Each of the subunits (or domains) includes multiple membrane-spanning amino acid sequences. Subunits self-assemble into a rosette, the center of which provides a pore through the membrane. The central pore is lined with hydrophilic amino acid residues, so that ions can pass from the aqueous environment at one end of the pore, along the internal, water-lined pore, and out to the aqueous environment at the far end. Thus ions never leave their preferred, polar environment in passing through the lipid barrier via protein channels.

I.2.3.1 Electrical analog for ionic channels

Electrophysiologists are interested in describing the function of ion channels as carriers of electrical current. The language of molecular biology or of biochemistry is of little help in such a description. Instead, electrophysiologists use the language of electricity in which the function of protein channels is described in terminology appropriate for electrical circuits. The protein-channel–free regions of cell membranes do not permit ion flow (they are infinite resistors, zero-value conductors) and, hence, do not appear in electrical circuits. Membrane proteins that form the ionic channels, on the other hand, are depicted in circuit diagrams as resistors. Because ions pass through each ion channel independently (that is, in parallel) the electrical analog of the membrane is a set of resistors (conductances) arranged in parallel. In such a parallel circuit the total conductance of the membrane is the sum of the conductances of the individual channels. Neurons typically have tens of thousands of the parallel conductances in their membranes.

I.2.3.2 Channel conductances

As we expect of proteins in general, ion channels formed by integral proteins have very specific properties and functions. Unlike pumps, which move ions up concentration gradients, channels can only facilitate the flow of ions down electrochemical gradients. No energy input is required for channels to function as ion carriers, although some energy expenditure may be required to control the state of the channels. The ease with which ions move through the pores formed by channel proteins is given by the channel conductance. Some typical single-channel conductance values are: potassium conductance (channels selective for K ions)—10 to 20 pS (picosiemens); sodium conductance (channels selective for sodium ions)—5 to 20 pS; the channel activated by the neurotransmitter, acetylcholine (ACh), at the neuromuscular junction—30 to 60 pS; and the gap-junction channel—about 150 pS.

I.2.3.3 Channel densities

Ionic channels are relatively rare proteins in cells, in part because they are largely confined to the lipid bilayers of cell membranes. The density of these proteins varies greatly. The densities for potassium channels in squid axons are about $60/\mu m^2$ and about $5/\mu m^2$ in frog muscle. The density of sodium channels in squid axons is greater, about $300/\mu m^2$. On the other hand, the density of leakage channel in the squid is very low, only about one channel per square micron. Finally, channel densities for acetylcholine receptors at the neuromuscular junction are very high, about $20{,}000/\mu m^2$.

I.2.4 PATCH-CLAMP RECORDING TECHNIQUE

The patch-clamp technique is a method whereby scientists can measure the state of individual ion channels embedded in lipid bilayers. To obtain such measurements, glass pipettes, usually about 1 mm in diameter, are heated and drawn to generate a tapering tube with a fine, hollow tip. The tip is then polished to generate a smooth opening with a diameter of about 1 μm. To provide a conducting path from the tip to the larger opening on the back end of the tube, the lumen of the pipette is filled with a salt solution. Depending on the specific patch-clamp configuration and experimental protocol, this solution might approximate the ionic concentration found in extracellular fluid or in the intracellular milieu. A silver wire inserted into the back end then completes the conducting path from the pipette tip to the electronic equipment that is used to measure membrane currents and to manipulate the membrane potential (fig. I.2-1).

Recordings from small patches of membrane are obtained by applying the tip of the patch electrode against the bare membrane of a cell and applying a small amount of suction to the pipette to draw the membrane against the smooth tip. With practice and luck, a very tight seal is formed between the membrane and the glass tip, effectively isolating the membrane patch encircled by the tip. The electrode can then be pulled away from the cell to generate an isolated inside-out membrane patch. In this configuration of the patch clamp, the cytoplasmic side of the patch of membrane is exposed to the saline solution surrounding the electrode while the external side of the membrane patch is exposed to the solution inside the electrode. Patch recordings can also be configured for the converse condition, an outside-out patch, in which the extracellular side of the cell membrane is exposed to bath solution. (Another configuration is the whole-cell patch clamp in which the membrane patch is not detached from the cell.)

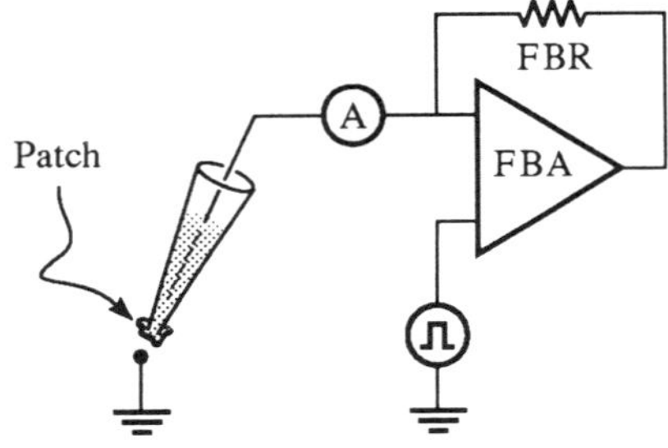

Figure I.2-1. Patch clamp apparatus. The pipette (size is greatly exaggerated) is attached to a small piece of membrane containing one or more ionic channels. The feedback amplifier (FBA) and resistor (FBR) function to hold the voltage across the membrane patch at a potential set by the experimenter (square pulse symbol). The current through the channels in the patch is measured by the ammeter (A).

Under favorable conditions only one or two ion channels will be located within the region of the patch bounded by the electrode tip. The properties of any channels in the patch can then be studied by varying the electrical potential across the patch and by varying the composition of the fluids on the two sides of the patch. For most experiments, the electronics (via a feedback amplifier) are configured to maintain a fixed potential difference across the patch. Measurements of the electrical current passing through the membrane then indicate whether ion channels present in the membrane are open or closed.

The recordings of patch-electrode currents have a very curious appearance; the current record jumps between levels, for example between zero and about 1 pA. In other words, the current is quantized. The zero level occurs when the channel is closed, whereas the second, fixed current level occurs when the channel is open. The size of the channel current is an indication of the single-channel conductance ($I = g * V$). Using the electrical analogy (figs. I.2-2a and I.2-2b), we can state that the membrane proteins that form ion channels act as resistors in series with a switch. Ions can flow through the resistor when the switch is closed (the channel permits passage of ions) but not when the switch is open (the channel is closed). We can apply Ohm's law to current flow through ion channels because experiments have demonstrated that, as in

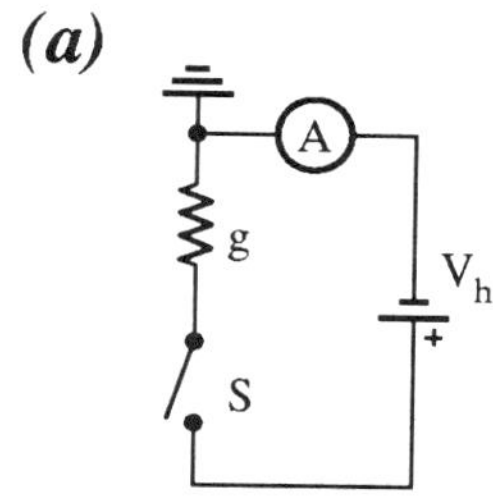

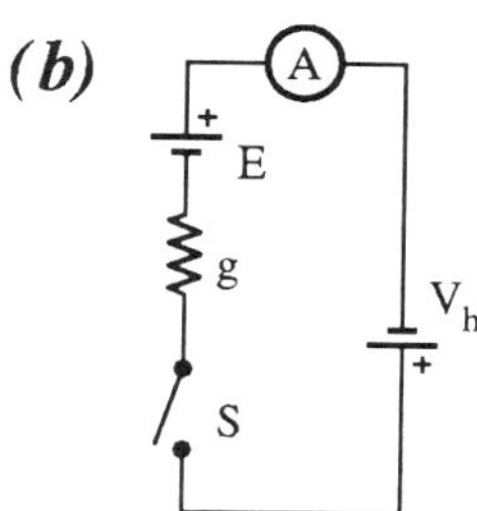

Figure. I.2-2. Electrical analogs of membrane patches. *a*) Analog for single ionic channel with equal concentrations of the permeant ions on each side of the patch. *b*) Analog for unequal concentrations of the permeant ions which generate a Nernst potential (E) symbolized by the battery. The potential across the patch is set by holding potential (V_h); currents are detected by the ammeter (A) each time switch S closes (i.e., when the channel opens).

electrical resistors, the magnitude of the current through open channels is proportional to the voltage (see modeling exercise Patch-1).

I.2.4.1 Voltage-gated channels

All membrane ion channels exhibit the open-closed behavior shown in the *NeuroDynamix* exercises. The state of the channel fluctuates randomly between these states. Specific openings and closings of ion channels cannot be predicted. (Strictly speaking additional, subconductance states, some of which can be observed electrophysiologically, also occur.) For some channels, the proportion of time that the channel remains open is controlled by the membrane potential; these are the voltage-activated channels, such as the sodium and potassium channels that generate nerve impulses (see section I.4). For other, ligand-activated channels, the opening time is controlled by the binding of neurotransmitter substances. Although the state of an individual channel at a specific time is not predictable, the average behavior of an individual channel and the average behavior of an ensemble of channels are completely predictable. As an example, a voltage-gated potassium channel (K channel) is closed most of the time when the membrane potential is more negative than −60 mV (measured from the inside with respect to the outside of the cell). This channel is mostly open when the potential is more positive than −40 mV. At any potential, however, the channels continue to fluctuate between the open and closed states (see modeling exercise Patch-2).

I.2.4.2 Reversal potential for voltage-gated channels

The amplitude of currents through ion channels is determined not only by the potential across the membrane patch, but also by the concentrations of the permeable ions on either side of the channel. As a simple example, we can see that no current can flow through the potassium channel if there are no free potassium ions nearby. In general, when ionic concentrations are very low, the analogy between ion channels and electrical resistors is not valid. A very common, less simple situation occurs whenever the concentration of the conducted ion is unequal on the two sides of the membrane (fig. I.2-2b). In this situation, the usual one in fact, currents flowing through ionic channels are described by a modified form of Ohm's law, one that takes into account the electrochemical gradient generated by the unequal concentrations. Such gradients give rise to a potential of their own (see section I.3). Taking this new potential into account, we can write Ohm's law for single, open ion channels as

$$I = g * (V_h - E_X), \tag{I.2-1}$$

where I is the (open) channel current, g is the channel conductance, V_h is the voltage (holding potential) applied by the patch-clamp electronics, and E_X is the electrochemical potential generated by the unequal distribution of ion X. This current fluctuates between some fixed value and zero as the channel protein changes its conformation between open and closed states. Note that no net current will pass through the channel when the holding potential is equal to the electrochemical potential, i.e., when $V_h = E_X$ in equation (I.2-1). As the holding potential is set to higher or lower values, the direction of the current reverses; thus, the membrane potential at which V_h equals E_X is called the "reversal potential" (see modeling exercise Patch-3).

Because of their distinctive molecular structures, ion channels exhibit specific, identifying electrical characteristics that are detectable with the patch-clamp technique. One such identifying characteristic is the sensitivity of channel gating to the holding potential voltage, as described above. Another is the selectivity of channels for specific ions; for although some membrane channels simply act as a nonselective pore through the lipid bilayer (gap junctions, for example), most channels are highly selective, permitting only specific ions to pass. That is why we can speak of potassium channels, sodium channels, and calcium channels; these permit passage of potassium, sodium, and calcium ions, respectively, while excluding other ions (see modeling exercise Patch-3).

I.2.4.3 Multiple channels

The single step-like changes of currents observed in patch-clamp recordings from cell membranes with a low density of ion channels are useful for studying channel function. The cell membranes of neurons usually include hundreds or thousands of channels. So also do membrane patches from membranes with a high channel density. Patch recordings from membranes that include multiple channels of the same type exhibit random, quantized current steps. These are the evidence that none, one, two, or more channels are open simultaneously (fig. I.2-3). In practice it is possible to be certain that multiple channels are present in a patch only if they are open simultaneously. When very many channels contribute to the recorded membrane current, the individual steps are no longer obvious; instead, one observes the ensemble average of all channels' currents. When thousands of channels are present, this average can be rather constant, obscuring the fundamental randomness of the currents contributed by individual channels (see modeling exercise Patch-4).

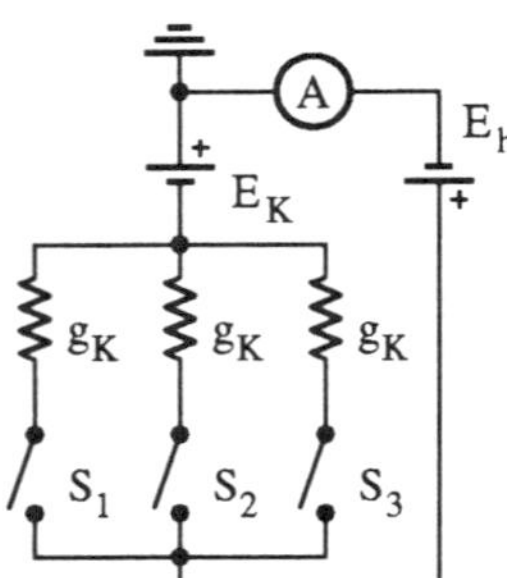

Figure I.2-3. Analog for membrane patch with multiple potassium channels. The switches open and close independently and randomly to generate a summed current that is detected by the ammeter (A).

I.2.4.4 Ligand-gated channels

Finally, there are ion channels whose behavior is controlled by the presence of neurotransmitter or neuromodulatory substances. The best understood is the nicotinic acetylcholine receptor channel (nAChR), first described at the neuromuscular junction. This channel molecule, a pentamere (five subunits), is primarily in a closed state until two molecules of acetylcholine (ACh) bind to its extracellular face. When activated by ACh, the ion channel permits both potassium and sodium ions to pass; the combined electrochemical potentials for sodium and potassium ions determines that the reversal potential for this receptor channel is about −10 mV. The channel is not sensitive to the membrane potential. Because two ACh molecules are required to open the channel, currents through the fraction of time that the acetylcholine receptors are open vary with the square of the ACh concentration (at low concentrations) (see modeling exercise Patch-5).

Modeling Exercises

NeuroDynamix Model: PATCH Exercises: Patch-1 to Patch-5

Introduction

This set of exercises is designed to demonstrate the fundamentals of the voltage-clamp technique for studying ion channels in cell membranes. The experiments introduce the types of records obtained from patch-clamp recordings and show how the recorded signals differ for each of the channel types. Figure I.2-1 illustrates the experimental situation that is simulated in these exercises. The setup includes a patch of membrane that is attached to the end of a glass pipette (tip diameter about 1 μm). The piece of membrane encircled by the glass tip is assumed to include one or more ion channels. The state of these channels—open or closed—is detected by the current that passes through the membrane when a constant electrical potential V_h is applied with the voltage-clamp apparatus, here schematically represented by the feedback amplifier circuit. In these exercises, as in the laboratory, the user can investigate the membrane patch by changing the holding potential, by altering the ionic composition of the solutions on the inside of the membrane or the pipette, and by applying a transmitter substance inside the pipette. These exercises include simulations of voltage-gated sodium, potassium, and calcium channels as well as simulations of non-voltage-gated chloride channels and nicotinic ACh channels.

Patch-1 Non-gated chloride channels

For this exercise PATCH is configured to reveal the currents through a single chloride channel. The electrical circuit modeled is that of a resistor (conductance) in series with a switch. This circuit is held at a fixed potential (voltage clamped) by an external voltage source. The current through the circuit is monitored by an ammeter (fig. I.2-2a). Two windows (fig. I.2-4) are open when you begin this exercise, a TIMESERIES window that graphs the current through the channel (*ICl*, in units of pA) and a PARAMETER window that shows that a single chloride channel is being studied (*Num-Cl* = 1).

The purpose of this exercise is to illustrate the simplest recording that can be obtained from patch-clamp recordings. The configuration simulates an isolated membrane patch with equal concentrations of chloride ions on the two sides; that is, *ECl* = 0. The potential difference across the patch (*Vhold*, in mV) may be set to fixed values to simulate the voltage-clamp electronics. Current passes through the ion channel only when the switch is closed. (To add realism to this simulation, noise has been superimposed onto the graph of the channel graph.)

Begin the exercise by clicking on the STOP/GO icon. Observe the upward current deflections in the TIMESERIES graph. The brief currents occur at

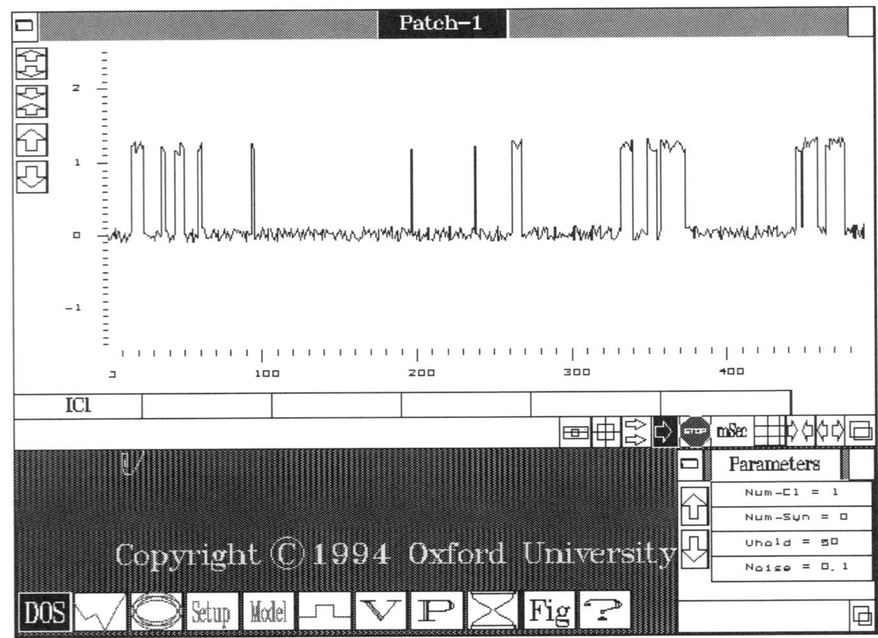

Figure I.2-4. Monitor display for exercise Patch-1

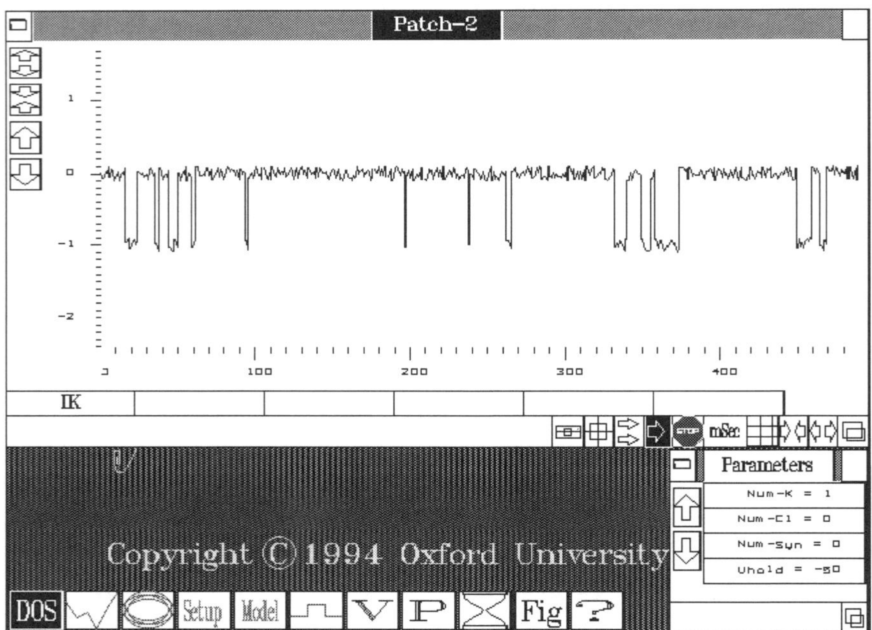

Figure I.2-5. Monitor display for exercise Patch-2

random intervals and last for a random duration but have nearly the same fixed positive value. These upward deflections represent chloride currents generated by the random opening and closing of a single channel. Note the amplitude of the chloride current, then use Ohm's law to calculate the channel conductance. (Note from the PARAMETER window that *Vhold* = 50 mV). Next, from several sweeps, estimate the fraction of time that the channel is open; this can be a rough estimate. Now set the amplitude of the holding potential (*Vhold*) to 70 mV. Again calculate both the channel conductance and the fraction of time that the channel is open. Change the holding potential to −50 mV and again calculate the value of channel conductance and the fractional open time. Are either of these parameters voltage dependent? The answer to this question tells you that this chloride channel is not activated by changes in membrane potential.

Patch-2 Patch-clamp recording of potassium channel current

For this exercise PATCH is configured to reveal the currents of a patch-clamp recording from a single potassium channel. The electrical circuit modeled is similar to Patch-1 except that the switch that controls whether current can pass through the resistor is sensitive to the applied voltage. The circuit again is held at a fixed potential (voltage clamped) by the external voltage source. The current through the circuit is monitored by an ammeter (fig. I.2-2b). Two windows (fig. I.2-5) are open when you begin this exercise, a TIMESERIES window that graphs the current through the channel (*IK*, in pA) and a PARAMETER window that shows that a single potassium channel is being studied (*Num-K* = 1).

The purpose of this exercise is to illustrate patch-clamp recordings from a more complex channel—one that is gated by the transmembrane potential. The configuration simulates an isolated membrane patch with equal concentrations of potassium ions on the two sides; therefore, *EK* = 0. The potential difference across the patch (*Vhold*, in mV) may be set to fixed values to simulate voltage-clamp electronics.

Begin the exercise by clicking on the STOP/GO icon. Observe the downward current deflections in the TIMESERIES graph; they are downward (negative) because *Vhold* is set initially to −50 mV. As we observed in recording from the chloride channel, the brief currents occur at random intervals and last for a random length of time; they all have nearly the same negative value. Note the size of the current. As in the preceding exercise, use Ohm's law to calculate the channel conductance (remember *Vhold* = −50 mV). Next, from several sweeps, estimate the fraction of time that the channel is open. Now increase the holding potential to −60 mV. Again calculate both the channel conductance and the fraction of time that the channel is open. Repeat your calculations with *Vhold* set to −70, −40, and −30 mV. Make a graph of *IK* versus *Vhold* for your five data points. Note that the points lie on a straight line whose slope is the channel conductance. Check your calculations by comparing the slope of your graph with the value for the potassium channel conductance given in the PARAMETER window (*gK*). Clearly the conductance of the open channel is not a function of membrane potential. Also graph your

Figure I.2-6. Monitor display for exercise Patch-3

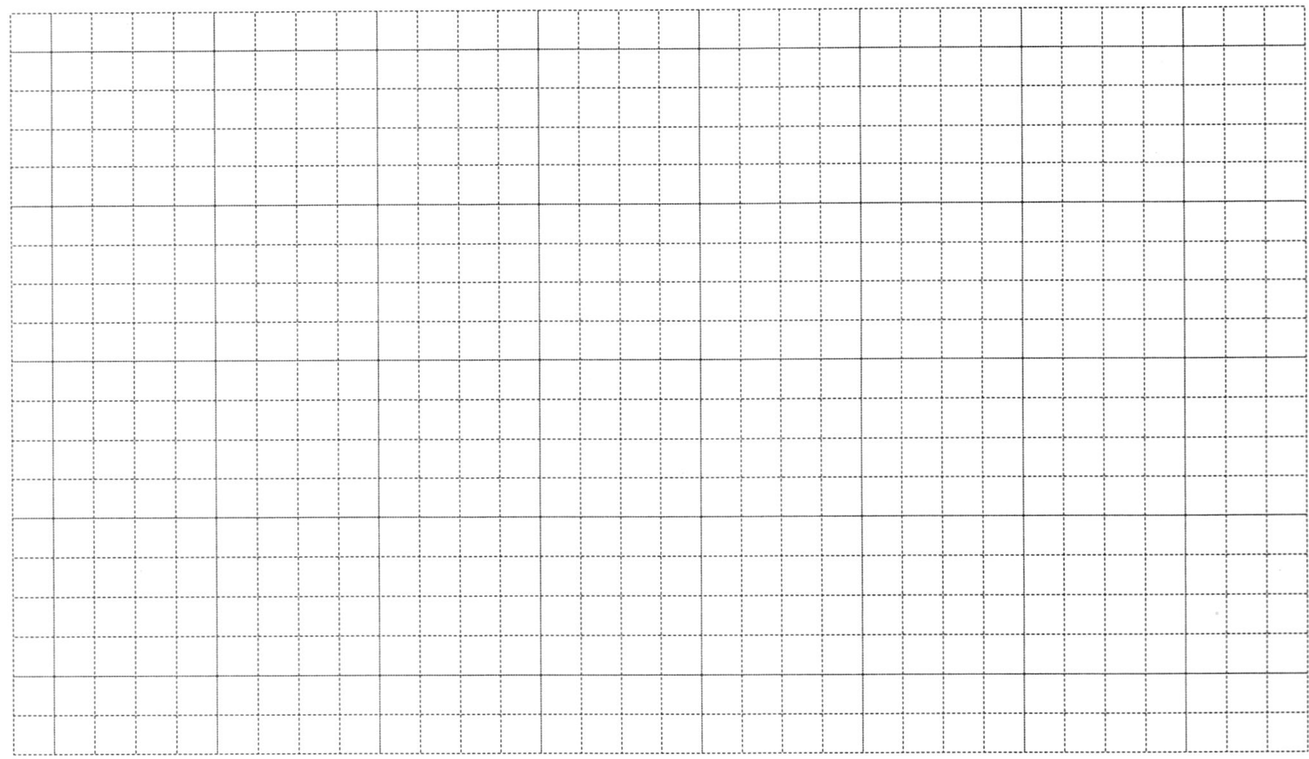

values of the fractional open times for the channel against *Vhold*. These points do not lie on a straight line. In fact, you can show that for *Vhold* values more negative than −70 mV the fractional open times are nearly zero and that for *Vhold* values more positive than −30 the channel is almost always open. The sigmoidal shape of your graph demonstrates the voltage sensitivity of this ion channel.

Patch-3 Identifying the channel type in patch-clamp recordings

For this exercise PATCH is configured to reveal the currents of a patch-clamp recording from a single sodium channel. The electrical circuit modeled is similar to Patch-2 except that a battery has been inserted in series with the switch and the resistor to simulate the Nernst potential for this membrane patch (fig. I.2-2b). Two windows (fig. I.2-6) are open when you begin this exercise, a TIMESERIES window that graphs the current through the channel (*INa*, in pA) and a PARAMETER window that shows that a single sodium channel is being studied (*Num-Na* = 1).

The purpose of this exercise is to illustrate how a sodium channel can be identified and described from patch-clamp recordings. The configuration simulates an isolated membrane patch with unequal concentrations of sodium ions on the two sides. For this simulation the concentrations are assumed to be those observed in intact cells; therefore, *ENa* = +55 (fig. I.2-2b). The potential difference across the complete circuit (*Vhold*) is again set by the (simulated) voltage-clamp electronics.

Begin the exercise by clicking on the STOP/GO icon. Observe the current deflections in the graph; they are downward and negative. In the intact neuron these currents would be inward and excitatory. One way of characterizing a channel is by experimentally determining the value of the Nernst potential for the membrane patch. To find this potential (namely, that potential at which no current passes through the channel—equilibrium), make a graph of the current through the channel for a range of holding potentials (*Vhold*) from −60 to +20 mV in 10 mV steps. Extrapolate a straight line through your data points to find the intersection of the line with the X-axis. The value of *Vhold* at this intersection point is the Nernst potential. The fact that this number is about +55 mV provides evidence that we are dealing with a sodium channel. Note that the slope of your line gives the channel conductance. Compare this value with the channel conductance value shown in the PARAMETER window (*gNa*—you will need to scroll down the parameter list). Your value should correspond to that of the sodium channel, providing further evidence that the patch does indeed have a single sodium channel.

Another technique for identifying the channel type is to alter the Nernst potential (accomplished by changing the ionic concentrations in the solution surrounding the membrane patch). Verify that your patch recording is of a sodium channel by changing *EK*, *ECl*, and *ENa* in the PARAMETER window (scroll down to these parameters). Which of these changes alters the channel

Figure I.2-7. Monitor display for exercise Patch-4

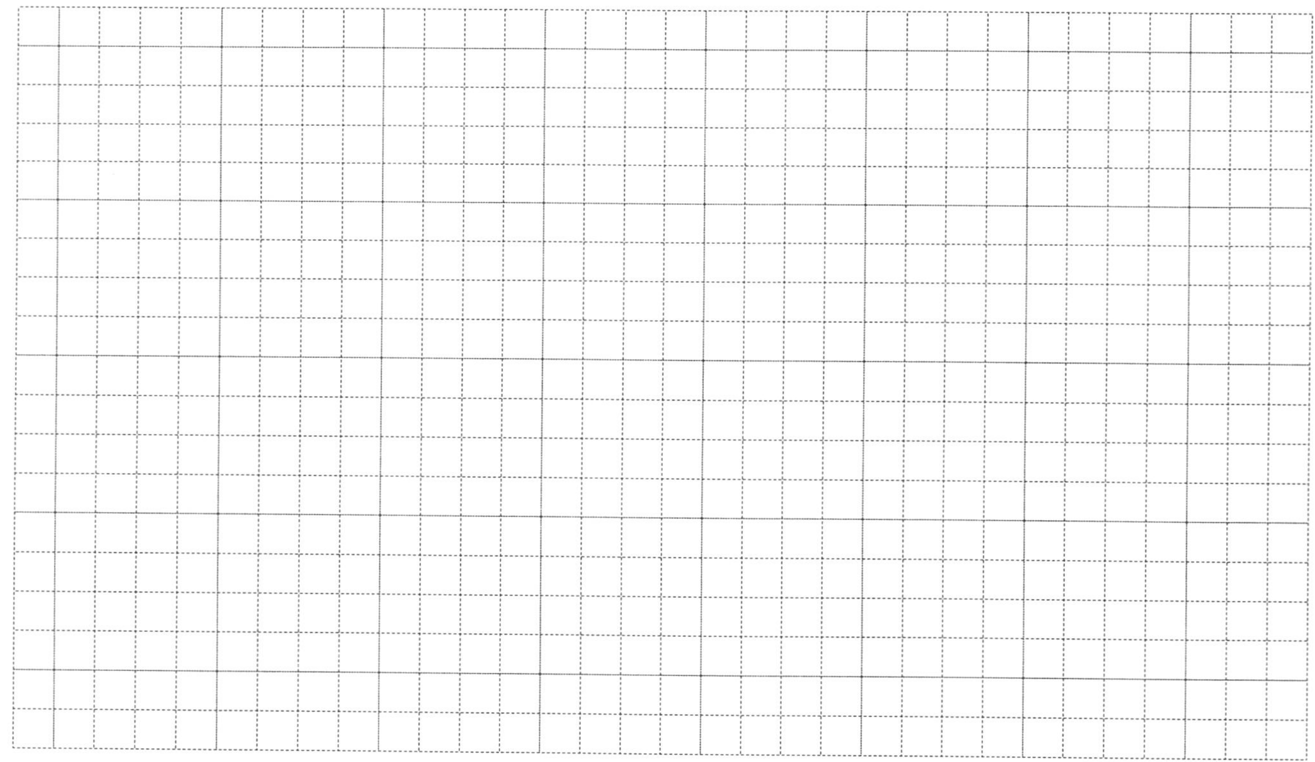

current? Note that although the amplitude of the current through the sodium channel is altered as *ENa* is changed, the open probability is unaffected. Why?

Patch-4 Membrane patch with multiple potassium channels

It often occurs in patch-clamp recordings that more than one channel is located under the electrode tip. Such multiple channels are simulated in this exercise. Here PATCH is configured to reveal the currents through several potassium channels (fig. I.2-3). The electrical circuit modeled is similar to Patch-3 except that there are several switches and resistors in parallel. Each one of these current paths acts completely independently. The circuit again is held at a fixed potential (voltage clamped) by the external voltage source. Here the current through the entire circuit (the sum of the current through the individual conductance channels) is monitored by the ammeter. The same two windows shown in the previous exercises are open when you begin this exercise (fig. I.2-7).

The purpose of this exercise is to illustrate patch-clamp recordings from multiple channels. The configuration simulates an isolated membrane patch with unequal concentrations of potassium ions on the two sides, as occurs in the intact cell; therefore, *EK* = −80 mV. The potential difference (*Vhold*) is set to −50 mV to ensure that the individual potassium channels are often open.

As you begin the exercise (click the STOP/GO icon) the number of potassium channels (*Num-K*) is set to 1. First observe the positive current deflections in the TIMESERIES graph; count the number of openings and closings as the trace moves from the left to the right of the screen. Now increase the number of channels in the membrane to two. Immediately you notice that the rate at which current pulses occur is doubled. Verify this by counting the new rate of channel openings, and compare this with your previous count. You will also note that occasionally there are current values that are twice those of the single channel size. The larger currents are due to the summed currents of two simultaneously open channels. The double-sized current deflections occur only on some sweeps of the trace across the graph. You could be fooled into thinking that there was only one channel (which opens and closes at a high frequency) if you did not wait long enough to observe these double-sized currents.

Increase the channel number to 3. The frequency of channel openings goes up again. Now, very occasionally, you see current steps that are three times the unit (single-channel) value. That is, the probability that all three channels open at once is very small. Again, do not be fooled into thinking that you have only two channels just because you do not observe the 3X currents!

Increase the number of channels to 100 by incrementing the parameter *Num-K* (use the PARAMETER MODIFICATION window) and rescale the ordinate to observe current values between 0 and 20 pA. Now the total potassium current (*IK*) is the sum of the currents through 100 randomly gated potassium channels. This is the type of record you might obtain from an intact cell that has only a small number of ion channels. What you see is a very noisy record, whose average deflection is equal to the fractional open time for the

Figure I.2-8. Monitor display for exercise Patch-5

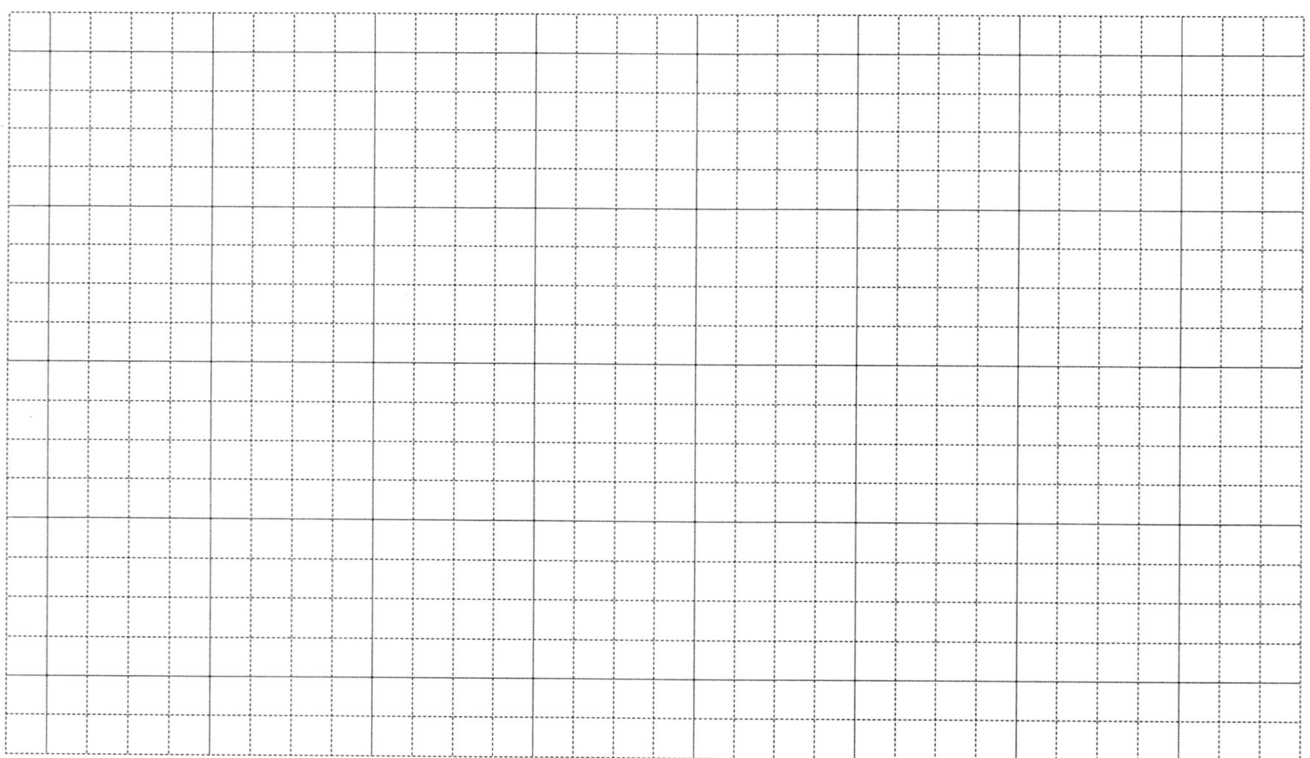

individual channels multiplied by the single-channel current and the number of channels. To verify this relationship, first divide the average current on the graph (with 100 channels) by 100. This is the mean current through each channel. Now compare this value with the directly calculated value for the average channel current. [Hint: Reduce the number of channels to 1; find the single-channel current; measure the fractional channel open time; and finally multiply these last two numbers to get the average channel current.] Your two measures of average channel current should be equal (except for measurement error).

Patch-5 Ligand-gated channels

For this exercise PATCH is configured so that you can examine the currents through a single nicotinic acetylcholine (ACh) sensitive channel. The electrical circuit modeled is much like that of Patch-3 except that this channel is gated by the concentration of a ligand rather than by membrane potential (fig. I.2-2b). This circuit is voltage clamped by an external voltage source. Three windows (fig. I.2-8) are open when you begin this exercise, a TIMESERIES window that graphs the current through the channel (*ISyn*, measured in pA), the PARAMETER window that shows that a single nACh channel is being studied (*NumSyn* = 1), and the STIMULATOR window that controls the concentration of ACh applied to the outside of the membrane patch.

The purpose of this exercise is to illustrate that ionic membrane channels can be gated by chemicals rather than by membrane potential. This isolated membrane patch has the normal concentrations of ions on the two sides; hence, the reversal potential for this synaptic channel is −10 mV (sodium and potassium ions pass nearly equally). The potential difference (*Vhold*) is set to −60 mV to mimic the normal potential difference across the cell membrane. The simulated concentration of ACh at the outside of the patch is controlled by the stimulator.

Begin the exercise by clicking on the STOP/GO icon. Initially the ACh concentration is set to zero (stimulator is off) and you observe very brief downward deflections due to exceedingly brief openings of the channel. Now apply 5 units of ACh by turning on the stimulator. The channel open times are now prolonged (although the amplitude of the current remains constant). Increase the concentration of ACh and notice that the fractional open time increases rapidly with concentration (make a graph). This rapid change reflects the fact that two ACh molecules must simultaneously bind to the nACh receptor in order to gate the channel open.

Verify that the reversal potential for this ligand-gated synaptic channel is indeed −10 mV by first setting the ACh concentration to 5 and then varying the holding potential (*Vhold*) from its initial value of −60 to +20 mV in 20 mV steps. Note the amplitude and sign of the current pulses for each value of *Vhold* and then make a graph of current amplitude versus *Vhold*. The intersection of a straight line through your points with the X-axis (the point at which the current amplitude is zero) is the value of the reversal potential. The fact that your points lie (nearly) on a straight line demonstrates that the single-channel conductance is a fixed value. Note that the fractional open time for this channel is not a function of the holding potential.

To get another view of the concentration dependence for synaptic-channel activation, increase the number of synaptic channels to 100 (*Num-Syn*). Scale the ordinate (*ISyn*) appropriately so that you can observe the summed current through all channels as you change the ACh concentration from 1 to 40. Notice that there is saturation (all channels are open) when the ACh concentration is large. (This simulation does not include receptor desensitization.)

Section I.3 Physical Basis for the Resting Potential

I.3.1 INTRODUCTION

Thanks to the pioneering experiments of the Italian physician Luigi Galvani (1737–1798), we have long known that electricity plays a vital role in the function of the nervous system. Galvani's experiments on frog sciatic nerve-muscle preparations demonstrated that electrical stimulation of nerves could cause muscle contraction. That these electrical signals are intrinsic to animal tissues was demonstrated by the German physician and electrophysiologist du Bois Reymond (1818–1896) who showed that cells have a "resting potential," the constant voltage across the cell membrane in the absence of stimulation, that disappears when the tissue is stimulated. The size of this resting potential was not determined in the nineteenth century because measuring devices were not adequate for detecting resting potential.

I.3.2 NERNST EQUATION

The physical basis for the resting potential was laid in 1889 by Nernst, a physical chemist. Nernst studied the potentials generated across inorganic membranes that are semipermeable, that is, membranes that are permeable to some ions but not to others. His theory established the relationship between ionic concentration and the electric potential across the membrane at equilibrium. We can understand the origins of Nernst's equation by considering a membrane permeable to one ionic species, X, with the two sides of the membrane bathed by solutions with differing concentrations of X. After equilibrium is established, the concentrations of X on side 1 is denoted by $[X]_1$ and on side 2 by $[X]_2$. At equilibrium there is no net flux across the membrane because no net work is required to transport a small number of ions, δn, from one side to the other. That is, the work of moving ions against the concentration gradient,

$$\delta Wc = \delta nRT \ln \{[X]_1/[X]_2\}, \tag{I.3-1}$$

is exactly balanced by the work of moving the ions through the electrical potential set up by the unequal ion distribution, namely,

$$\delta We = \delta nZF(V_2 - V_1). \tag{I.3-2}$$

In these equations

R (8.31 joules/degree mole) is the gas constant;
T (293°K at 20°C) is the absolute temperature in Kelvin degrees;
ln (2.303 $\log_{10}$) is the natural log;
F (96,400 Coulomb/mole) is Faraday's constant;
Z is the valence of the permeant ions; and
$V_2 - V_1$ is the electrical potential across the membrane.

To simplify the terminology we set the electrical potential difference, $V_2 - V_1$, to E. Because $\delta Wc = \delta We$ at equilibrium, we have

$$\delta nRT \ln \{[X]_1/[X]_2\} = \delta nZF(V_2 - V_1). \tag{I.3-3}$$

By cancelling the δn terms and rearranging equation (I.3-3) we are left with

$$E_X = (RT/ZF) \ln \{[X]_1/[X]_2\}. \tag{I.3-4}$$

Equation (I.3-4), the Nernst equation, describes the potential difference across a membrane that is permeant to only one type of ion. It is important to emphasize that this equation holds only at equilibrium, when there is no net movement of ions and that it holds only if there is only a single, permeant ion species. However, the value of the Nernst potential can be calculated whenever the intracellular and extracellular concentrations of ions are known.

The physiologist Bernstein (1839–1917) realized that the Nernst equation may be applied to neurons if side 1 of the Nernst membrane represents the extracellular space and if side 2 represents the cell interior. Because electrophysiologists arbitrarily set the potential of the extracellular fluid to zero, E_X, the Nernst potential describes the intracellular potential for the specific (and unrealistic) situation in which only one ionic species can cross the cell membrane. To consider a specific example, suppose that some cell has a plasma membrane that is permeable only to potassium ions. We apply the Nernst equation (with $Z = +1$) and predict that the membrane potential of this cell will be

$$E_K = (RT/F) \ln \{[K^+]_{out}/[K^+]_{in}\}. \tag{I.3-5}$$

At room temperature (20°C) the term RT/F reduces to 25.2 * 10^{-3} V (= 25.2 mV). We make a final transformation from the natural log to the base-10 log to get

$$E_K = 58 \log \{[K^+]_{out}/[K^+]_{in}\} \text{ (mV)}. \quad \text{(I.3-6)}$$

As listed in table I.3-1, $[K^+]_{out}$, the extracellular potassium concentration, is relatively low (5 mM in mammalian muscle tissue); whereas $[K^+]_{in}$, the intracellular concentration, is much higher (140 mM). With these values we can calculate the Nernst potential for potassium ions in muscle cells (recorded at a temperature of 20°C) as

$$E_K = 58 \log (5/140) = -84 \text{ mV}. \quad \text{(I.3-7)}$$

The Nernst equation applies to *any* ion, so we may also calculate the Nernst potential (i.e., that potential which would occur if only a single ionic species were membrane permeable) for sodium ions for muscle cells as

$$E_{Na} = 58 \log \{[Na^+]_{out}/[Na^+]_{in}\} = 58 \log (145/10) = +67 \text{ mV}. \quad \text{(I.3-8)}$$

Similar calculations give the values of Nernst potentials for other ionic concentrations and additional ions (see table I.3-2). Note that these values hold only at 20°C and that the value of Z, the valence, must be correctly incorporated into the calculations. For temperatures other than 20°C it is necessary to use the full, explicit Nernst equation, such as equation (I.3-5) for the potassium equilibrium potential, rather than simplified equations, such as equation (I.3-6) (see modeling exercises Soma-1 and Soma-2).

I.3.3 THE HODGKIN-HUXLEY-KATZ MODEL FOR THE RESTING CELL MEMBRANE POTENTIAL

We now combine our picture of cell membranes composed of a lipid insulator in which channel proteins act as conductance pathways (section I.2) with electrical concepts (section I.1). (See figure I.3-1a for the conceptual experimental scheme for studying simple membrane parameters.) In the macroscopic, parallel conductance model of Hodgkin, Huxley, and Katz

Table I.3-1 Representative values of ionic concentrations (in mM)

	Na^+	K^+	Cl^-	Ca^{2+}
Mammal muscle cells				
Intracellular	10	140	2	0.0001
Extracellular	145	5	77	2
Squid giant axons				
Intracellular	50	400	60	0.0001
Extracellular	460	20	560	10

Table I.3-2 Nernst potentials (approximate values at 20°C)

	E_{Na}	E_K	E_{Cl}	E_{Ca}
Mammalian cells	+67 mV	−84 mV	−60 mV	+125 mV
Squid axons	+55 mV	−75 mV	−60 mV	+125 mV

(fig. I.3-1b and section I.4) the ensemble of conductance channels for each type of permeant ion is represented by a single resistor. For studies of neuronal electrophysiology the conductances for chloride, calcium, sodium, and potassium ions are of primary importance. Our model of the resting cell membrane includes four conductances, which selectively carry chloride, calcium, sodium, and potassium ions through the membrane. Each of the four types of ion channels has an associated equilibrium potential given by the Nernst equation. When the membrane is set to the equilibrium potential for any one of these conductances, there will be no net current through that type of ionic conductance. The Nernst potential for each ionic species is incorporated into the parallel conductance model by placing a battery with potential E, the Nernst potential, in series with each conductance. (Strictly speaking, the value of each battery is set to the absolute value of the Nernst potential, with the battery polarity indicated by appropriately orienting the battery symbol in the circuit diagram. Also, because we are simulating the resting potential, we are ignoring the membrane capacitance, which is a fifth circuit component, in parallel with the four current paths.) The electrical model in fig. I.3-1b can represent either a unit area of membrane or an entire cell (if the membrane potential is the same throughout the cell), provided that the values of resistors are scaled appropriately.

The parallel conductance model is highly useful for electrophysiologists because all of the parameter values can be measured or computed. For the moment we will assume that the current carried by the membrane pump I_P is zero, that it therefore can be ignored, and that no current is generated by the constant-current source. We can now easily derive the expression for membrane potential in this model using Ohm's law and Kirchhoff's node rule. First, we can see by inspection (the node rule) that the sum of the ionic currents is equal to the current from the constant-current source (i.e., equal to zero), so that

$$I_{Cl} + I_{Ca} + I_{Na} + I_K = 0, \qquad \text{(I.3-9)}$$

where the four terms are the currents for chloride, calcium, sodium, and potassium ions, respectively. The ionic currents are described by Ohm's law ($I = g * V$; remember that the conductance, $g = 1/R$, is the useful terminology when circuits have parallel conductances). The term V is the voltage across a conductance g. Because the batteries are in series with the conductances, the potential across each conductance is the difference between the membrane potential and the battery potential, so that

$$I_{Cl} = g_{Cl} * (V_m - E_{Cl}), \qquad \text{(1.3-10)}$$

$$I_{Ca} = g_{Ca} * (V_m - E_{Ca}), \qquad \text{(1.3-11)}$$

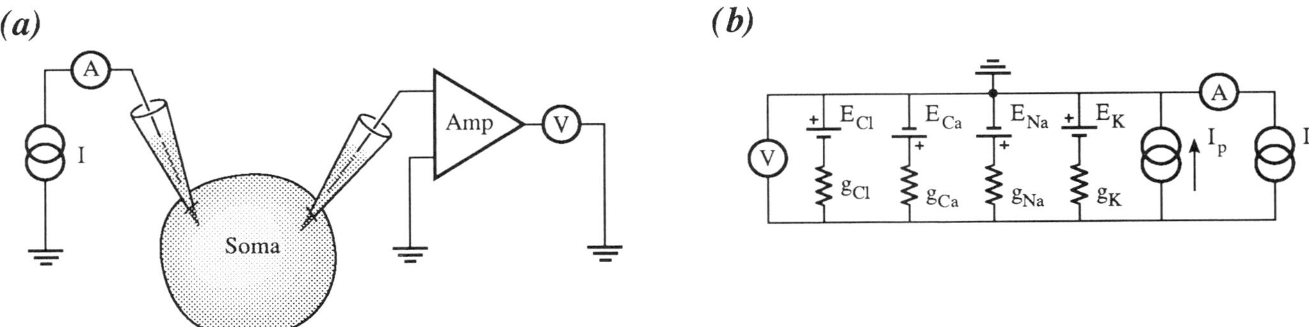

Figure I.3-1. SOMA model. *a*) Diagram of the experimental preparation modeled by SOMA. Two electrodes, which pass current and measure membrane potential (amplifier, A, and voltmeter, V, respectively) are simulated. *b*) Circuit diagram for the SOMA model. *I* is the current applied by the experimenter.

$$I_{Na} = g_{Na} * (V_m - E_{Na}), \text{ and} \quad (1.3\text{-}12)$$

$$I_K = g_K * (V_m - E_K). \quad (1.3\text{-}13)$$

In these equations the terms in the parentheses, the voltages across the conductances in figure I.3-b, are the electrical forces that drive ions through membrane channels. Moreover, these terms determine the sign of the currents because conductances are always positive. Thus, when the driving force is negative, the ionic current is negative (inward, into the cell); conversely, when the driving force is positive, the current is positive (outward). To illustrate these points some illustrative values for these electrical driving forces are given below. (Values are calculated for Nernst potentials in squid axons, assuming that $V_m = -60$ mV).

K^+: $[-60 - (-75)] = +15$ mV (current is outward)
Na^+: $[-60 - (+55)] = -115$ mV (current is inward)
Cl^-: $[-60 - (-60)] = 0$ mV (no current)
Ca^{2+}: $[-60 - (+125)] = -185$ mV (current is inward)

By definition, the resting potential of a cell is the constant membrane voltage when cells are not excited by external stimulation, by synaptic inputs, or by their endogenous, time- and voltage-dependent conductances (see sections I.4, I.5, and I.6). Under these quiescent conditions, the membrane potential can be derived by substituting equations (I.3-10 to I.3-13) into equation (I.3-9) to yield

$$g_{Cl} * (V_m - E_{Cl}) + g_{Ca} * (V_m - E_{Ca}) + g_{Na} * (V_m - E_{Na}) + g_K * (V_m - E_K) = 0. \quad (I.3\text{-}14)$$

Through the use of a little algebra, we can solve this equation for V_m to get

$$V_m = E_r = \frac{g_{Cl} * E_{Cl} + g_{Ca} * E_{Ca} + g_{Na} * E_{Na} + g_K * E_K}{g_{Cl} + g_{Ca} + g_{Na} + g_K}, \quad (I.3\text{-}15)$$

where E_r is the value of the membrane potential in the resting state. A useful way of viewing this equation is to realize that the resting potential of a cell is the sum of the equilibrium (Nernst) potentials for all permeant ions weighted by the relative size of the electrical conductance for that ion. Thus if g_T is defined as $g_{Cl} + g_{Ca} + g_{Na} + g_K$, the resting potential becomes

$$E_r = E_{Cl} * g_{Cl}/g_T + E_{Ca} * g_{Ca}/g_T + E_{Na} * g_{Na}/g_T + E_K * g_K/g_T. \quad \text{(I.3-16)}$$

These electrical conductances can often be determined experimentally with relative ease (see modeling exercise Soma-3).

I.3.4 SOURCE OF MODEL PARAMETERS

The parameters that describe the "resting" electrophysiology of specific cells can be derived through a series of experimental manipulations. First, the total resting conductance can be obtained through *I-V* curves. To make this measurement, long-lasting current pulses are injected into the cell while the change in membrane potential is recorded. For detailed measurements, a graded series of current pulses is applied. The membrane potential excursions caused by these pulses are then plotted against the injected current to generate an *I-V* graph. When the *I-V* relationship is a straight line, its slope is the total membrane resistance; the reciprocal of the slope is the membrane conductance ($g_T = 1/R$). Values of specific ionic conductances are obtained by using channel blockers (or by removing ions) to eliminate all but one ionic current. Under these conditions, the equilibrium (Nernst) potentials are given the values of membrane potential—in the absence of external currents. If there are voltage- and time-dependent conductances, much greater effort (see section I.4) is required to determine the specific conductance values of the conductances in the parallel conductance model (see modeling exercise Soma-4).

I.3.5 THE ELECTROGENIC SODIUM/POTASSIUM PUMP

The disparity in the ionic concentrations between the cytoplasm and extracellular fluid is generated by a sodium/potassium ATPase—that is, by a membrane pump that splits ATP to transport potassium ions into cells and sodium ions out. This pump exhibits rather standard enzyme-substrate characteristics, such as saturation. There is a maximum pumping rate; and this pumping rate depends on the concentration of the substrates, sodium and potassium ions. The process of transport involves energy-requiring conformational changes in the pump protein. This protein acts as an asymmetrical antiport, with two potassium ions transported inward through the membrane and three sodium ions transported outward for each ATP molecule split.

Because of the asymmetry in ionic transport, the sodium/potassium pump generates an electrical current, I_p, out of the cell. Flowing across the resting conductances of the cell, this current generates a voltage (I_p/g_T) that adds to the resting potential given in equation (I.3-16). So, a more complete equation for the potential of a cell is given by

$$E_r = E_{Cl} * g_{Cl}/g_T + E_{Ca} * g_{Ca}/g_T + E_{Na} * g_{Na}/g_T + E_K * g_K/g_T - I_p/g_T + I/g_T. \qquad \text{(I.3-17)}$$

We have included an additional term, I/g_T, to describe membrane potential changes imposed through the injection of external current. Note that because the pump current is a positive, outward current it reduces the number of intracellular positive ions and drives the membrane potential to more negative values.

The most important physiological control of pump rate is the intracellular concentration of sodium ions. For inactive cells the pump rate is usually small and the contribution of the pump to membrane potential is minor. (It can be very large in plant cells.) However, following intense stimulation, which loads neurons with sodium ions, the pump rate can increase considerably and hyperpolarize a cell by many millivolts. For completeness, it should be mentioned that pump function also requires the presence of potassium ions in the extracellular fluid. The concentration of potassium ions usually is nearly constant, but when reduced experimentally, the rate of ionic pumping decreases. Pump action is also blocked by ouabain, a cardiac glycoside, which competes with potassium ion binding (see modeling exercise Soma-5).

A note on convention—the current injected into a cell by the experimenter is assigned a positive value if positive ions are injected into the cell. This positive current acts to depolarize the cell.

Figure I.3-2. Monitor display for exercise Soma-1

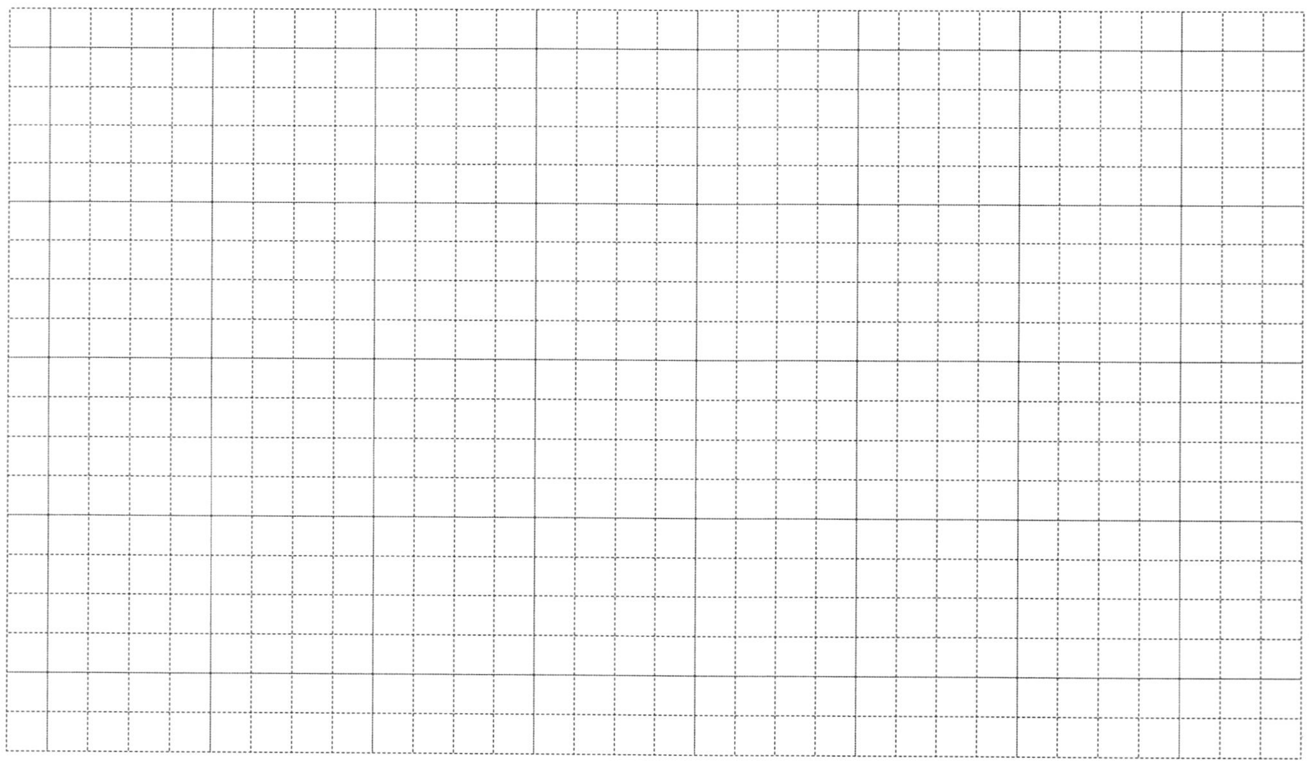

Modeling Exercises

NeuroDynamix Model: SOMA Exercises: Soma-1 to Soma-5

Introduction

This set of exercises is designed to illustrate the fundamental factors that generate the transmembrane potential. They demonstrate explicitly the properties of the electrical analog model of Hodgkin/Huxley/Katz. The experimental preparation envisioned for these exercises is that of an isolated neuronal soma (cell body) whose membrane potential is monitored with one intracellular electrode and into which current may be injected with another electrode (fig. I.3-1a). With this model preparation we can investigate the origins of the membrane potential by changing the current injected into the cell, by manipulating the ionic composition of the intracellular space, and by adjusting the extracellular ionic environment. For these exercises only a few, simple channel types and an electrogenic sodium pump are included. The soma is assumed to include a very large number of channels for each type of conductance. All measurements are of macroscopic currents.

Soma-1 Nernst potential as a function of ionic concentrations

For this exercise SOMA is configured to explore the nature of the Nernst equation applied to a cell whose membrane is permeable only to potassium ions (see equation I.3-5). This equation predicts that the membrane potential will be proportional to the absolute temperature and to the log of the external and internal potassium concentrations. Three windows (fig. I.3-2) are open when you begin this exercise, a TIMESERIES window that graphs the value of the Nernst potential, E_K, as a function of time, a PHASEPLANE window that graphs E_K against the log of the external potassium concentration (*LCKOut*), and the PARAMETER window. The purpose of this exercise is to illustrate the importance of ionic concentration in determining the Nernst potential.

To begin this exercise, first open the PARAMETER MODIFICATION window for *CKOut* (measured in mM) so that you can manipulate the external potassium concentration; then click on the STOP/GO icon. You will observe that *EK*, graphed in the upper window, has a fixed value that is determined by the initial external and internal potassium concentrations, 140 and 5 mM, respectively (temperature is set to 20°C). What is the Nernst potential, E_K? Now double the external potassium concentration and again measure E_K. Notice first that increasing the concentration of external potassium ions *decreases* the value of E_K. Also note that although the concentration was doubled, the change in E_K is much less than twofold. What is the concentration at which the potential will be zero? Test your guess. If you set the external potassium concentration to 140 mM (equal to that on the inside), you observed that the Nernst potential did

Figure I.3-3. Monitor display for exercise Soma-2

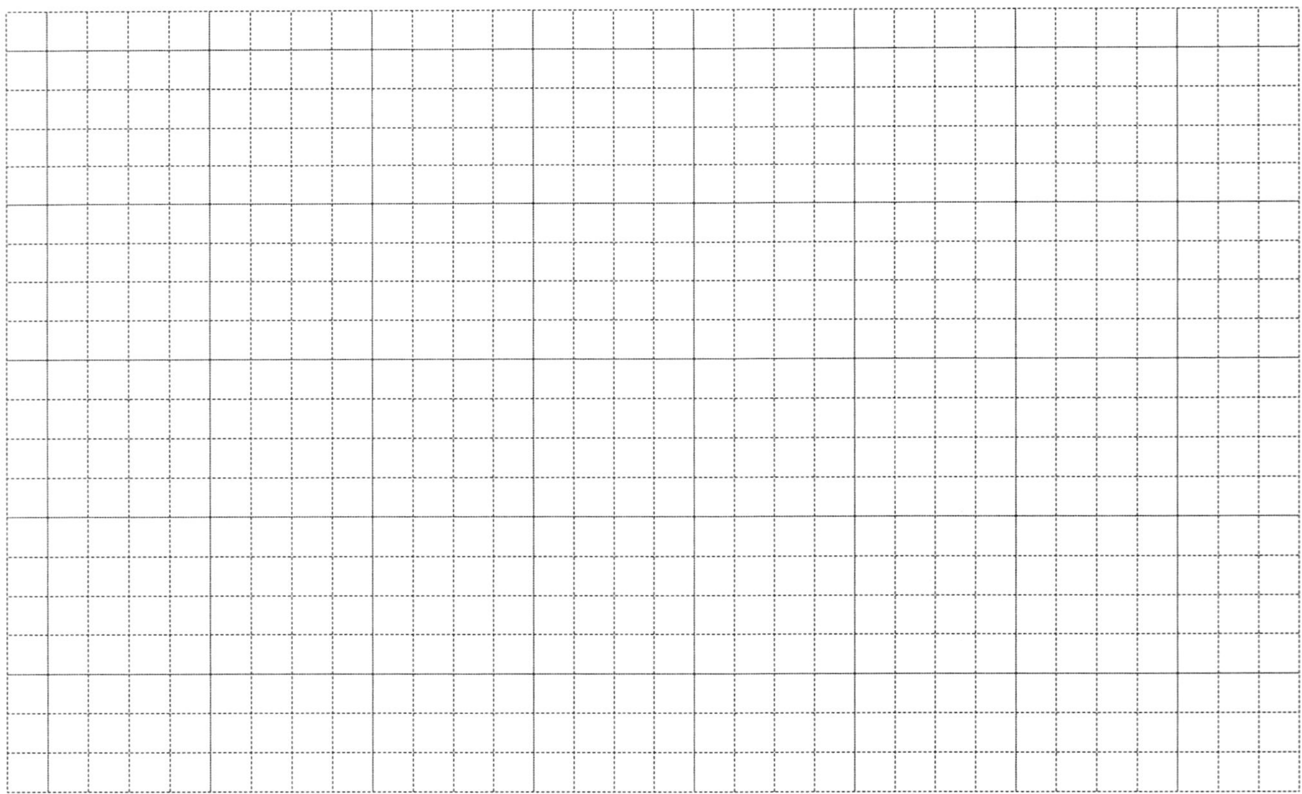

indeed go to zero.

Use the lower window to verify that there is a log relationship between E_K and the external potassium concentration. First set *CKOut* to 140 mM; then clear the PHASEPLANE window. Now decrease *CKOut* in a series of steps (try 140, 100, 10, 1, 0.1 mM). You will see from the lower window that a straight line is formed, demonstrating that E_K is indeed linearly related to the log of the extracellular potassium concentration. By measuring the slope of this line you can verify for yourself that each tenfold change in potassium concentration (a one-unit change on this log scale) increases or decreases the Nernst potential by 58 mV. You can construct similar curves for E_{Cl}, E_{Na}, and E_{Ca} by selecting the appropriate variables to plot on the PHASEPLANE window ~~and by setting all conductances, except for the ion of choice, to zero.~~ (Stop graphing while you replace variables, and note that you will need to rescale the ordinate for the latter two values.) Why are the slopes not 58 for E_{Cl} and E_{Ca}?

Soma-2 Nernst potential as a function of temperature

For this exercise SOMA is configured as above, but with the ordinate on the TIMESERIES window greatly expanded (fig. I.3-3) to show alterations in E_K with much greater precision. The purpose of this exercise is to illustrate that the Nernst potential is sensitive, however only slightly, to the temperature of the experimental preparation.

To begin this exercise, first open the PARAMETER MODIFICATION window for *Temp*; notice that the units are in degree Celsius. When graphing begins (click on the STOP/GO icon) the potassium concentrations are at the normal values and the temperature is 20°C. The Nernst potential (upper window) is near −84 mV. Reduce the temperature by 20 degrees and observe that E_K is reduced by about 5 mV. An increase of about 5 mV occurs if you increase the temperature from 20°C to that of mammals, about 37°C. (Therefore, the normal value of E_K for mammalian muscles is actually about −89 mV.)

Soma-3 The Hodgkin-Huxley-Katz model for the resting potential

For this exercise SOMA is configured with three parallel conductances—for sodium, potassium, and chloride ions (fig. I.3-1b; calcium conductance removed). The ionic concentrations, and hence the Nernst potentials, in this exercise reflect those found in the squid giant axon; however, the conductances associated with these ions are chosen merely to illustrate the dependence of membrane potential on these ions without any particular cell in mind. The conductances and currents in this exercise therefore are in arbitrary units; only the relative values are of importance. Three windows are open when you begin this exercise: a TIMESERIES window that graphs the values of V_m, E_{Na}, E_K, and E_{Cl}; a second TIMESERIES window that graphs the ionic sodium, potassium, and chloride currents; and the PARAMETER window (fig. I.3-4).

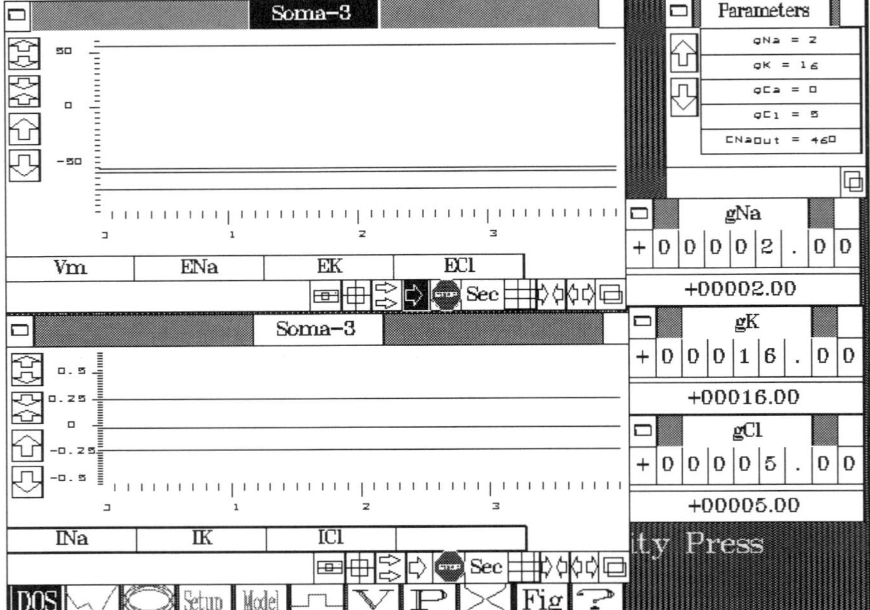

Figure I.3-4. Monitor display for exercise Soma-3

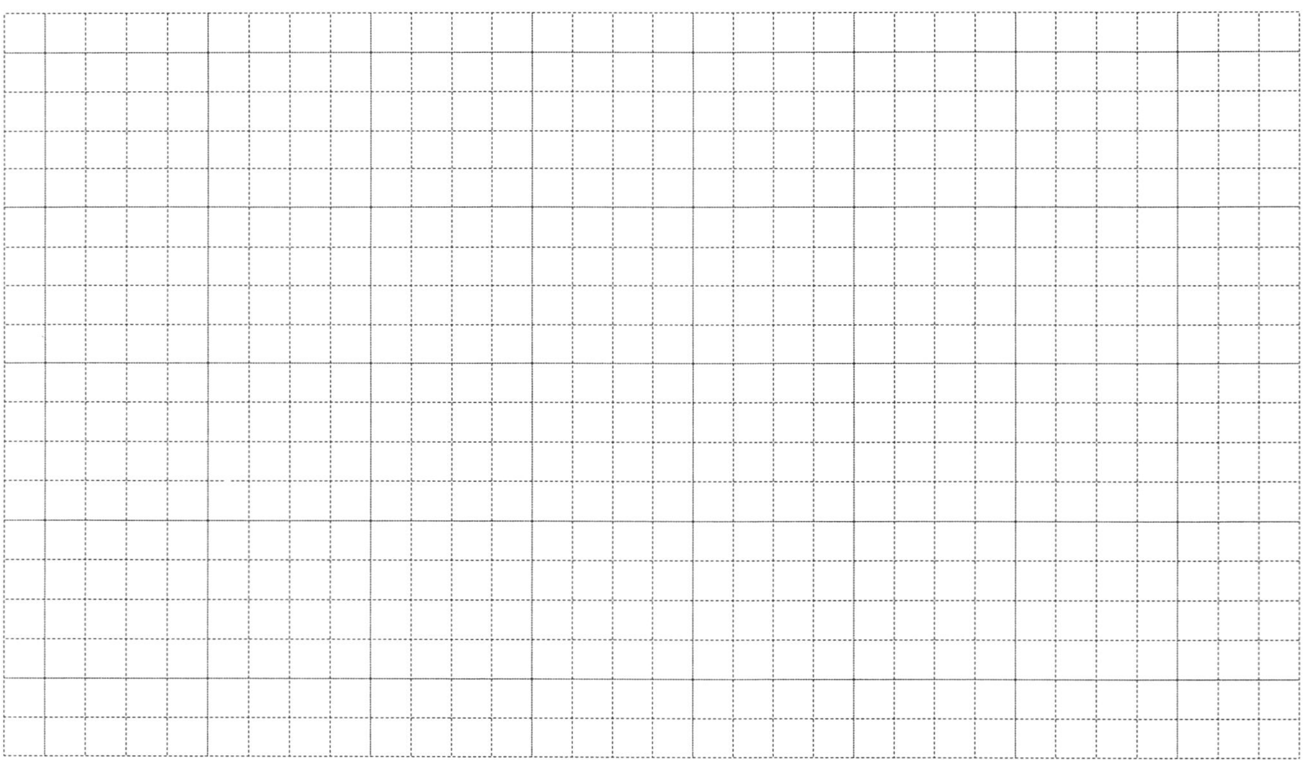

The purpose of this exercise is to illustrate how the cell membrane potential is determined by the relative values of the chloride, sodium, and potassium conductances and by the equilibrium potentials of these ions.

To begin this exercise, open the PARAMETER MODIFICATION windows for the sodium, potassium, and chloride conductances. When you begin graphing (click the mouse on the STOP/GO icon), you will notice in the upper window that *Vm*, the membrane potential, lies just below E_{Cl} and between the sodium and potassium Nernst potentials. You can see in the lower TIME-SERIES window that the system is not in equilibrium; there are a net negative (inward) current for sodium ions, a net positive (outward) current for potassium ions, and a small, negative chloride current. The sum of these three currents (the only currents in the circuit) must equal zero. Because the chloride current is so small (why?), the sodium and potassium currents are nearly equal in magnitude (although opposite in sign).

To gain some appreciation of how the membrane potential is influenced by the relative membrane conductances of the three ions, raise and lower the conductances for sodium, potassium, and chloride ions. Notice that if any two conductances are set to zero, the membrane potential goes to the value of the Nernst potential of the third ion. Conversely, if the conductance for any ion is made very large, V_m assumes a value very near the Nernst potential for that ion. In particular, notice that when the sodium conductance is increased by a factor of 50 (to 100), the membrane potential approaches +30 mV, near the peak value observed for nerve impulses (see section I.4). Contrary to the strong influence of sodium conductance on membrane potential, alterations in the conductance for chloride ions raise or lower the membrane potential only slightly (assuming that the sodium and potassium conductances are near their initial values).

Soma-4 Experiments to find the values of membrane conductances and Nernst potentials

For this exercise SOMA is configured as for exercise Soma-3, with three parallel conductances—for sodium, potassium, and chloride ions (fig. I.3-1b). In this exercise the ionic concentrations, and hence the Nernst potentials, reflect those found in the mammalian cells. The conductances are in units of nS, and the currents are in nA to reflect values that might be found in recording from mammalian cells. Four windows (fig. I.3-5) are open when you begin this exercise: a TIMESERIES window that graphs V_m; a PHASEPLANE window that graphs V_m against the stimulus current (*IStim*); the PARAMETER window; and the STIMULATOR window. The purpose of this exercise is to illustrate the procedures by which the total membrane conductance, the values of conductances for individual ions, and the Nernst potentials for these conductances can be determined experimentally. These latter values can then be used to predict the resting potential in the Hodgkin-Huxley-Katz model of cell membrane potential.

Begin this exercise by clicking the mouse on the STOP/GO icon to show that the resting potential of this model cell, with its three conductances at their

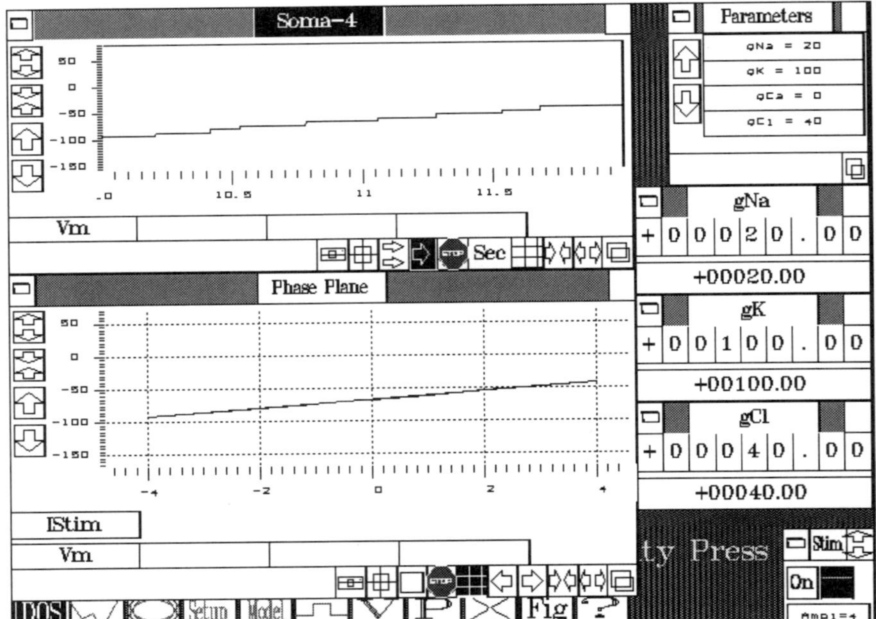

Figure I.3-5. Monitor display for exercise Soma-4

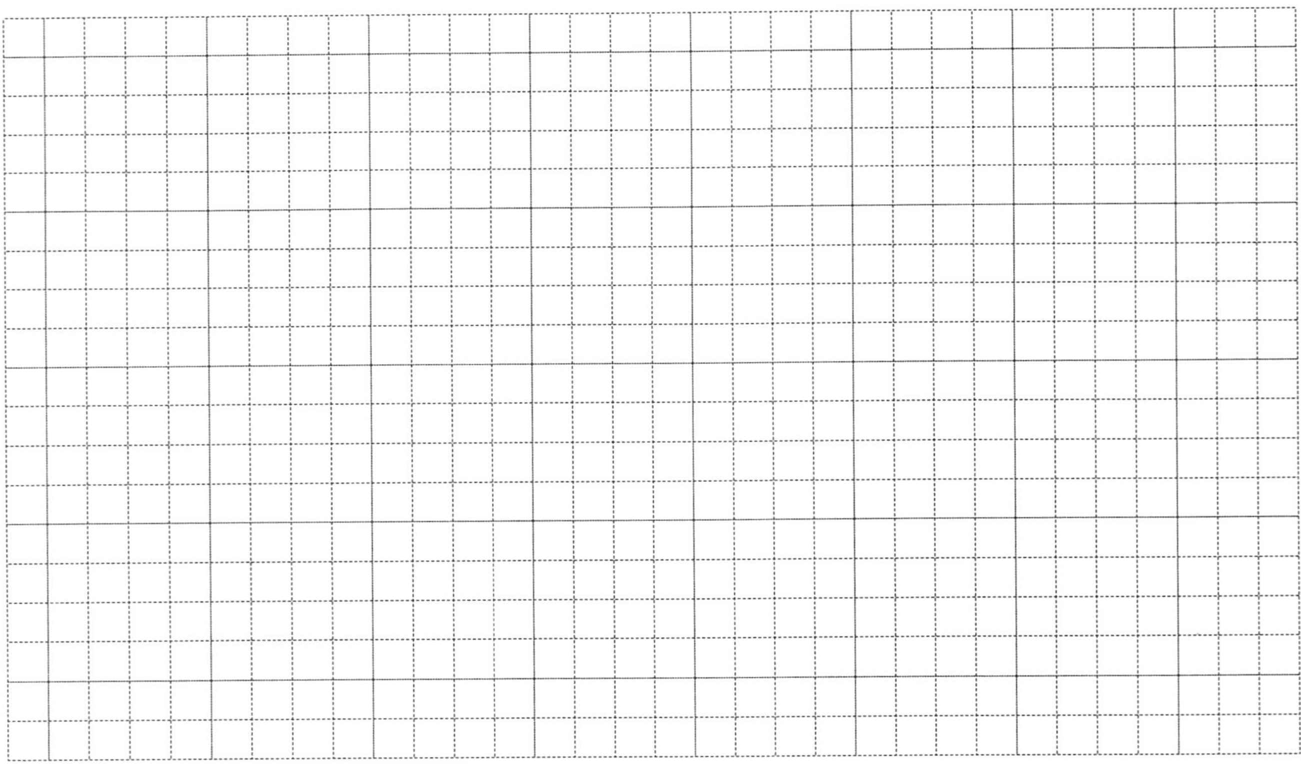

normal values, is constant at about −63 mV. Now open PARAMETER MODIFICATION windows for the sodium, potassium, and chloride conductances and set the values of the sodium and chloride conductances to zero. Inject current into the model cell by changing the amplitude of the stimulator stepwise from 0 to +4 nA to −4 nA. Note the small size of the membrane excursions induced by these currents in the upper, TIMESERIES graph, and note the small slope of the *I-V* graph in the lower, PHASEPLANE window. Recalling that the slope of this graph is equal to the value of the resistance (and the reciprocal of the conductance), you can conclude that the potassium conductance of this cell is large. The potassium equilibrium potential is the membrane potential when the stimulus amplitude is zero. (What is its value?) Set the current amplitude and potassium conductance to zero. Increase the sodium conductance to 20 nS and repeat. Then inject current pulses and repeat the observations. How does this new graph differ from the previous one? Repeat with the chloride conductance set to 40 nS (others set to zero). Calculate the resting potential for this model cell by substituting your calculated (from the reciprocals of the slopes) values of g_K, g_{Na}, and g_{Cl} and your measured values of the corresponding Nernst potentials into equation (I.3-16) with $g_{Ca} = 0$. Compare this value with the actual resting potential obtained from the TIMESERIES graph with $g_K = 100$ nS, $g_{Na} = 20$ nS, and $g_{Cl} = 40$ nS.

Soma-5 Role of the electrogenic sodium pump in setting membrane potential

For this exercise SOMA is configured as for exercise Soma-4 but with the addition of an electrogenic sodium pump (fig. I.3-1b). This simulation best describes a neuronal soma (with a cell volume of 0.1 picoliter) in cell culture. For this simulation the internal sodium ion concentration is not fixed; rather it is determined by the relative values of the sodium influx through the sodium conductance and the efflux provided by pump action. (The relationship between current and flux is $I = F * J$, where I is charge movement in Coulombs/s, J is particle movement in mole/s, and F is the Faraday constant = 96,400 Coulombs/mole.) When the current is expressed in nA the corresponding flux is in nmole/s. Four windows (fig. I.3-6) are open when you begin this exercise: two TIMESERIES windows (one is a graph of *Vm*, the other graphs *INa*, *IPump*, and *CNaIn*—the intracellular sodium ion concentration) the PARAMETER window; and the STIMULATOR window.

The purposes of this exercise are to illustrate the contribution of the electrogenic pump to the resting potential and to demonstrate the role on intracellular sodium ions in setting the pump rate. For the latter demonstration, the stimulator has been configured to inject sodium ions into the simulated neuron without directly causing membrane depolarization. (This situation is realized in real experiments by penetrating cells with two electrodes, one containing a high concentration of sodium ions, and then passing a current between the two electrodes.) Note that in this simulation the *effective* pump current (the component that hyperpolarizes the membrane) is set to 1/3 of the total pump current in order to mimic the characteristic coupling between sodium

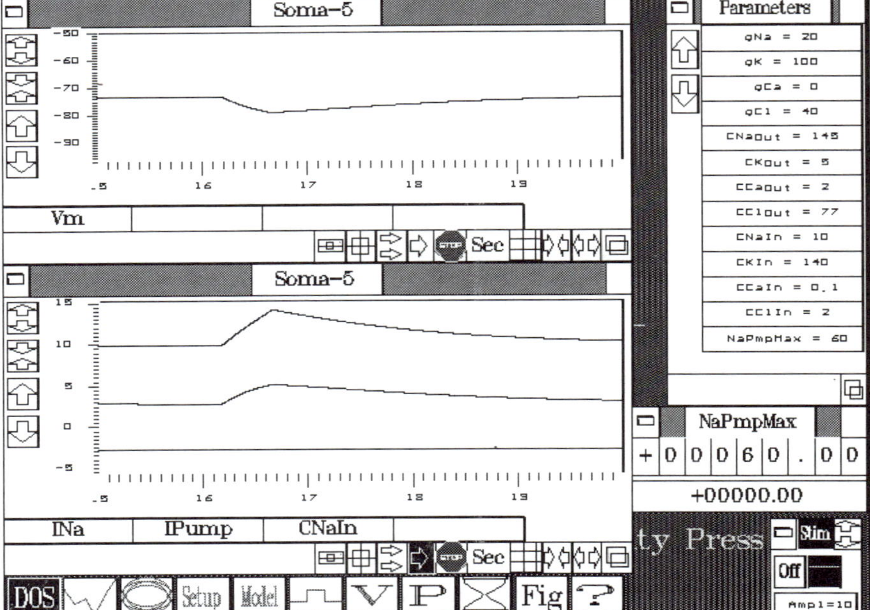

Figure I.3-6. Monitor display for exercise Soma-5

efflux and potassium influx in the Na pump. However, the potassium influx is not directly demonstrated here.

At the beginning of this exercise (click the STOP/GO icon) the pump is turned off (*NaPumpMax* = 0). You will see that the membrane resting potential is about −67 mV and that the concentration of intracellular sodium ions is near 10 mM. Now open the PARAMETER MODIFICATION window for *NaPumpMax* and set the value of this parameter to 60 (units are femtomole/s). At this pump rate the intracellular concentration of sodium ions is unaltered (still 10 mM), nevertheless the electrogenic nature of the sodium pump hyperpolarizes the cell membrane by about 6 mV. Change the maximum pump rate. You will see that increasing the rate causes an immediate hyperpolarization; but as the intracellular sodium concentration is reduced by the increased pump activity, the hyperpolarization is much reduced. These effects reverse when the pump rate is increased. Can you explain why?

You should note that, although the maximum pump rate is fixed by the *NaPumpMax* parameter, the actual pumping rate depends in a sigmoidal manner on the intracellular sodium concentration. Thus when the internal concentration is high the pump is much more active than when the concentration is low. To demonstrate this phenomenon, set *NaPumpMax* back to 60 and wait for the steady state to be re-established. Now inject sodium ions into the cell by turning the stimulator on (+10 nA is a good value). You will first observe that the intracellular sodium concentration is elevated (of course!) by the injection of sodium ions. The elevated sodium concentration activates the pump (*IPump* trace on the lower graph), which hyperpolarizes the membrane. This hyperpolarization is observed in real neurons after similar injection of sodium ions or, more physiologically relevant, after prolonged stimulation increases intracellular sodium concentrations.

Section I.4 Basis of the Nerve Impulse

I.4.1 INTRODUCTION

The conduction of rapid signals by neurons over distances greater than about 1 mm occurs almost exclusively by electrical impulses. This chapter presents the fundamental concepts about the mechanisms that underlie the generation and propagation of these impulses. After a brief historical summary to indicate our understanding of these electrical events in 1950, the remainder of the chapter describes the voltage-clamp experiments on the giant axon of the squid performed by A. L. Hodgkin and A. F. Huxley and published in 1952, . The results and the conclusions furnished by these definitive experiments remain the major source of our conception concerning the electrical nature of nerve signals. Even after a span of forty years, Hodgkin and Huxley's explanation for the origin of nerve impulses in squid axon applies, with minor modification, to all long-distance signaling in all nervous systems. The Hodgkin-Huxley experiments are presented here in some detail both because of their central importance for electrophysiology and because they were carried out in a remarkably logical sequence. The Hodgkin-Huxley study on the squid axon still serves as a model of scientific research at its best.

I.4.2 THE HISTORICAL SETTING

The rate of progress in all areas of biology depends critically on finding appropriate animal or plant experimental preparations. For neurobiology one source of difficulty in learning about the details of neuronal physiology is that most nerve cells are minuscule functional units (about 10 μm in diameter). Although massed electrical activity can be measured in such large, homogenous tissues as muscles in which many fibers act in concert, the electrical activity of nerve cells, and of muscles, is intrinsic to individual cells. Thus, whereas many informative experiments were carried out in the early twentieth century with extracellular electrodes and with pseudo intracellular recordings (obtained by crushing a part of the tissue), a full description of electrophysiology awaited the experimental capacity to obtain electrical records from the *inside* of cells. Only with electrodes placed simultaneously on both sides of cell membranes, on the *inside* and on the outside, was it possible to measure directly and accurately the transmembrane potential. Measurements of such intracellular potentials became feasible in the middle of the twentieth century when J. Z. Young rediscovered the giant axons of squids in 1936. These axons, sometimes mistaken for blood vessels, can measure up to 1 mm in diameter. Their large size permitted, for the first time, the use of electrical probes such as glass pipettes or silver wires to measure the internal electrical potentials in neurons.

An impressive array of information concerning the electrical properties of nervous systems was available in the middle of the twentieth century. Among the relevant information known to Hodgkin and Huxley was the following. First, it was understood that the conduction of information in nerves and muscles is via exceedingly brief electrical events known as nerve (and muscle) impulses (also "action potentials"). Because the conduction velocity of nerve impulses is much less than the conduction of electricity in wires (1–100 m/s, rather than 3×10^8 m/s), it was clear that the nature of these two types of electrical processes are very different. It was also known that differences between potassium (and chloride) ion concentrations within and surrounding cells maintain an electrical potential, the resting membrane potential (section I.3), across cell membranes. The movement of sodium ions was known to play a critical role in generating the nerve impulse. In fact, as early as 1902, E. Overton had suggested that nerve impulses are accompanied by an exchange of sodium and potassium ions across the cell membrane. That the presence of sodium ions in the extracellular medium is essential for nerve impulse activity was known through the experiments of Hodgkin and Katz. These scientists showed that reduction of external sodium concentrations reduced both the amplitude and the rate of rise of the nerve impulse. Furthermore, Hodgkin and Huxley had shown by experiments on the squid axon in the late 1930s that there is a reversal, from negative to positive, of the cell membrane potential when impulses occur. The peak amplitude of these impulses is about 100 mV (0.1 V). Finally, it was known through experiments by H. J. Curtis and K. S. Cole that the electrical conductance of the cell membrane undergoes a fortyfold increase coincident with the occurrence of nerve impulses. This increased conductance (that is, decreased resistance) of the membrane to the flow of ions during the nerve impulse was predicted already by Bernstein. Together with the information about the electrical nature of the nerve impulse summarized here, this impulse-associated increase in conductance led Hodgkin and Huxley to propose the theory that nerve impulses are generated by transient, sequential increases in cell membrane conductance to sodium and potassium ions. They designed their seminal experiments, which won them the Nobel Prize in 1963, to test this theory (see modeling exercise Axon-1).

The technical developments that gave rise to fast electronics during the first half of the twentieth century provided neurophysiologists with the capability to measure accurately the exceedingly rapid changes in voltage that are recorded as nerve and muscle impulses. These developments were extended by Cole in the late 1940s. He invented a means, through the use of a feedback amplifier to control membrane currents, of clamping the electrical potential across the cell membrane in the squid giant axon. The idea here was to supply just the right amount of current to maintain the membrane potential at a constant level. The voltage clamp senses the membrane potential, compares this potential with the potential value set by the experimenter, and then automatically, via a feedback amplifier, generates a current that is injected into the cell. This current acts continuously to minimize the difference between the "command" potential and the actual membrane potential (fig. I.4-1). The voltage-clamp technique and its derivative, the patch clamp, is used today in most studies on the biophysics of the cell membrane.

In performing their pioneering experiments Hodgkin and Huxley took full advantage of the large size of the squid axon. They dissected out pieces of axon of measured length and diameter and then inserted two silver wires into the inside, one to measure the internal potential, the other to pass current. They also took advantage of the fast voltage-clamp electronics developed by Cole to clamp the potential across the axon membrane. As described above, by using the voltage clamp they could set the axon potential to selected, fixed values and then measure the clamp current required to achieve and maintain these values. With this experimental preparation they set out to examine in detail the membrane currents that are elicited as the membrane potential of the squid axon is changed rapidly from one value to another.

I.4.3 THE MEMBRANE ANALOG

In sections I.1 to I.3 we introduced the analog electrical circuit as a means of describing the electrical properties of cell membranes. This equivalent circuit, the parallel conductance model for the cell membrane, first proposed by Hodgkin and Huxley as an analog for the giant axon of the squid, is shown in figure I.4-2. The parallel conductance model, which represents neuronal membrane with uniform internal and external potentials, was appropriate for the squid preparation because the squid axon was "space clamped." That is, the internal potential of the axon was maintained at a uniform, spatially homogeneous value by the two silver wires extended along the length of the

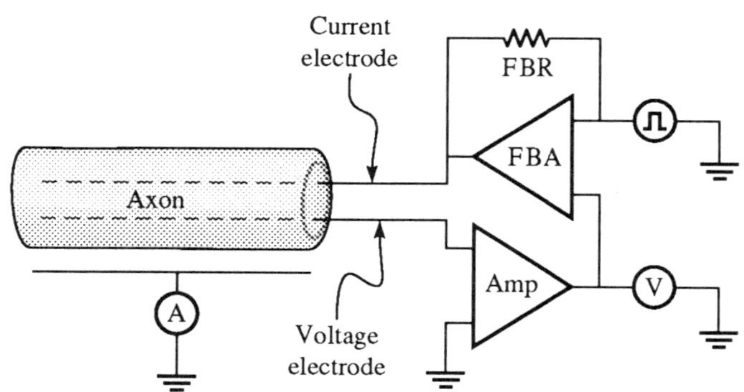

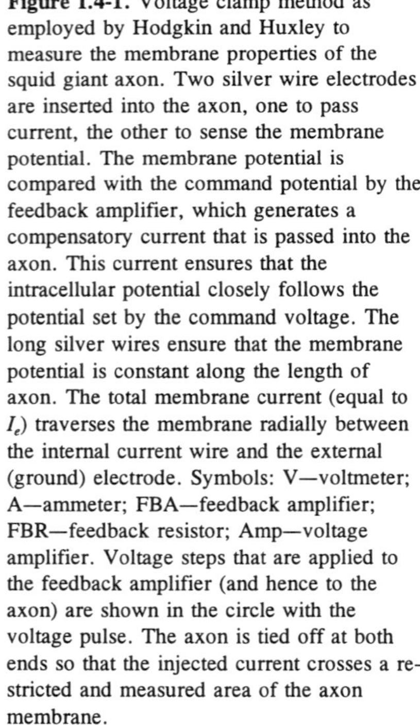

Figure I.4-1. Voltage clamp method as employed by Hodgkin and Huxley to measure the membrane properties of the squid giant axon. Two silver wire electrodes are inserted into the axon, one to pass current, the other to sense the membrane potential. The membrane potential is compared with the command potential by the feedback amplifier, which generates a compensatory current that is passed into the axon. This current ensures that the intracellular potential closely follows the potential set by the command voltage. The long silver wires ensure that the membrane potential is constant along the length of axon. The total membrane current (equal to I_e) traverses the membrane radially between the internal current wire and the external (ground) electrode. Symbols: V—voltmeter; A—ammeter; FBA—feedback amplifier; FBR—feedback resistor; Amp—voltage amplifier. Voltage steps that are applied to the feedback amplifier (and hence to the axon) are shown in the circle with the voltage pulse. The axon is tied off at both ends so that the injected current crosses a restricted and measured area of the axon membrane.

axon. The extended current wire ensured that only radial currents, through the membrane, were generated. In our previous presentation of the parallel conductance model the conductances, which provide the paths for current to pass through the lipid bilayer, had fixed values. For individual channels, the states were either on or off (section I.2); for channel ensembles conductance was unaffected by membrane potential (section I.3). In the analog circuit for the squid axon there are three conductances in parallel with the membrane capacitor. Two of the conductances, the sodium conductance (labelled g_{Na}) and the potassium conductance (labelled g_K) are variable. The third conductance (labelled g_{Leak}) is constant. Two quantities can be measured directly in the physical realization of this analog circuit: the total current between points labelled "inside" and "outside" and the voltage (potential) between these points. These quantities can be measured also in the actual squid axon. By means of the specific analogy between a measured length of squid axon of known diameter and the electrical analog circuit, current and voltage values obtained from the physiological squid preparation could be incorporated directly into the electrical model.

The Hodgkin-Huxley model incorporates a total of four current paths across the membrane (fig. I.4-2). The two current paths designated by variable resistors (potentiometers) correspond, in modern terminology, to two types of membrane channels, permeable specifically to sodium ions and potassium ions. The efficacy of these paths for carrying current is described by the conductances g_{Na} and g_K, respectively. A third current path, designated by the fixed conductance, corresponds to an ion channel that is less selective; hence, it acts as a "leak" with conductance g_{Leak} $(= g_l)$ and carries a "leakage" current that is a mixture of ions. The current in each of these three paths is determined by the voltage difference between the inside and outside and by the battery voltage associated with each conductance.

The fourth and analytically most complex current path in the analog model is through the capacitor, C_m, which is the circuit analog of the excellent capacitor formed by the lipid bilayer and the conducting fluids inside and outside of the cell. Recall from section I.1 that no charge carriers actually pass through a capacitor; instead, charges simply pile up or are depleted on the conducting plates. In the physiological membrane, this means that no ions pass through the membrane capacitance; rather, they pile up or are depleted next to the membrane. Therefore, this capacity (displacement) current is non-ionic.

The currents through the four paths of the analog circuit are given by the following equations (see sections I.1 to I.3):

$$I_{Na} = g_{Na}(V_m - E_{Na}), \qquad \text{(I.4-1)}$$

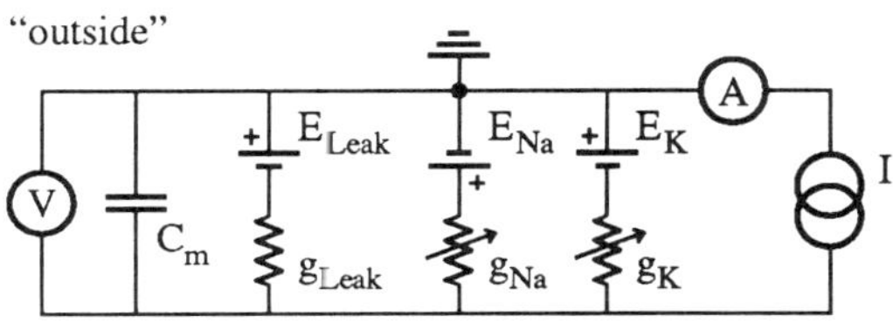

Figure I.4-2. Parallel conductance model for squid axon. This is the equivalent circuit for the piece of axon depicted in fig. I.4-1. The parallel conductance circuit is at the heart of the Hodgkin-Huxley-Katz model for the electrical function of excitable cell membrane. Symbols: C_m—membrane capacitance; E_{Na}—battery potential analog for the sodium reversal potential; E_K—battery potential analog for the potassium reversal potential; E_{Leak}—battery potential analog for the leakage reversal potential; g_{Na}—resistor analog for membrane conductance for sodium ions; g_K—resistor analog for membrane conductance for potassium ions; g_{Leak}— resistor analog for membrane conductance for leakage ions. I is the current applied by the voltage clamp or by the experimenter. The upper and lower terminals represent the outside and inside of the membrane, respectively.

$$I_K = g_K(V_m - E_K), \text{ and} \qquad \text{(I.4-2)}$$

$$I_l = g_l(V_m - E_l) \qquad \text{(I.4-3)}$$

for the ionic currents and

$$I_C = C_m \mathrm{d}V_m/\mathrm{d}t \qquad \text{(I.4-4)}$$

for the displacement current in the capacitor path. In these equations V_m, the voltage between inside and outside, and the I terms are variables under the control of the experimenter. The equilibrium potentials, E_{Na}, E_K, E_l (= E_{Leak}), are fixed battery potentials (determined by intra- and extracellular ionic concentrations); the capacity of the membrane C_m also is fixed. The conductances are variable functions of time and of membrane potential.

The current passed between the inside and outside of the circuit, namely the total current passed by the experimenter, I, is equal to the sum of the ionic and capacity currents; hence,

$$I = I_{Na} + I_K + I_l + C_m \mathrm{d}V_m/\mathrm{d}t. \qquad \text{(I.4-5)}$$

Equations (I.4-1 to I.4-5) describe the membrane analog circuit (fig. I.4-2) completely. If the relationships between currents and voltages are determined experimentally, then the conductances, the battery potentials, and the capacity can be calculated. It was precisely these relationships that Hodgkin and Huxley obtained with their voltage-clamp experiments on the giant axon. As a consequence they then could calculate the changes in membrane conductance as a function of time and compute the potential in the axon. (Please note: The notation and sign convention used here are those currently in use; they differ from those employed in the original scientific papers published by Hodgkin and Huxley in 1952.)

I.4.4 THE VOLTAGE CLAMP

As can be seen from equation (I.4-5), the total current across the membrane includes the displacement current through the capacitor, which depends explicitly on the rate at which the membrane potential changes. The extraordinary usefulness of the voltage-clamp technique is that it provides a means of eliminating this non-ionic current. With the voltage clamp, the membrane potential is stepped rapidly from one constant value to another. While the potential is constant, $\mathrm{d}V_m/\mathrm{d}t = 0$ and, consequently, $I_C = 0$. During the transition between different potential values $\mathrm{d}V_m/\mathrm{d}t$ is obviously not zero; in fact, it is very large during rapid transitions. With modern, fast electronics the duration of the transition between differing membrane potential values is exceedingly brief (about 20 μsec); hence, the large surge of capacity current associated with the transients between potentials is completed before changes in membrane conductance occur. These capacity currents are removed in the data collection process and usually are not displayed in illustrations of voltage-clamp data. Therefore the data derived from voltage-clamp experiments are measures of the much more prolonged total ionic current. Obviously, one trick for obtaining good voltage-clamp measurements is to make the transition interval as brief as possible.

For a thorough understanding of the voltage-clamp methodology it is necessary to study the procedures and to learn some terms. As the name implies, when the voltage clamp is applied to a tissue, the membrane potential is held at a fixed value, the commanded potential. The membrane potential maintained by the clamp in the absence of commanded step changes is the "holding potential." Often, though by no means always, this potential is set near the normal resting potential of the cell. Changes from this holding potential are in steps in which the potential is commanded to go to some new level for a predetermined interval. This "step" potential is described either by specifying the value of the new, commanded potential or as the magnitude of the step, i.e., as the difference between the holding potential and the new potential. One might say that the membrane is stepped from a holding potential of −60 mV to a new potential of −20 mV or that, equivalently, the potential is altered by a step of +40 mV. Although conceptually simple, the voltage-clamp procedure for measuring transmembrane currents includes some complexities. The step to a new potential might, for example be preceded by a "conditioning" step. Similarly, there might be interposed a second step between the primary step and the return step to the holding potential. Because most membrane conductances are complex functions of membrane potential, many voltage-clamp experiments include a series of steps, separated by return steps to the holding potential. In this way, the voltage dependence of currents (consequently of the underlying conductances) can be determined. Similarly, because of the time dependence of conductances, voltage-clamp experiments often include a series of constant-voltage steps of increasing duration. Such experiments generate series of current traces that are then superimposed graphically.

I.4.5 VOLTAGE-CLAMP EXPERIMENTS ON THE SQUID GIANT AXON

Remember that the aim of Hodgkin and Huxley was to characterize the changes in membrane conductances that give rise to the axon impulse. Recall also that by using the voltage clamp they could impose membrane potential values on the axon and then determine the amplitude and time course of the *total* membrane current. With the elimination of the capacity current, there remain three currents that contribute to the total current and whose underlying conductances must be characterized. Of these, the leakage conductance is the simplest because it is constant, independent of membrane potential. Hodgkin and Huxley determined the value of this conductance first. They noted that sodium and potassium conductances are nearly zero at the normal resting (holding) potential and are turned on or "activated" only when the membrane potential is stepped to some less negative (depolarized) potential. By stepping the membrane potential to successively more negative (hyperpolarized) values, they obtained a series of values for the leakage currents; then, using equation (I.4-3), they calculated the leakage conductance. Following subtraction of the leakage current from the total ionic current, the current remaining is the sum of only two ionic currents, the sodium current and the potassium current. This subtraction procedure works even for membrane depolarizations because the leaking conductance is constant; hence, the magnitude of the leakage current can be predicted (equation [I.4.3]) for any membrane potential step. This small current, positive for depolarizing and negative for hyperpolarizing steps from the resting potential, is then subtracted from the measured total ionic current.

I.4.5.1 Combined sodium and potassium currents

As can be inferred from figure I.4-2, the membrane currents that remain after the capacity and leakage currents are removed or subtracted are those due to the movements of sodium and potassium ions. The conductances that underlie these currents, because they generate the complex waveform of the nerve impulse, can be expected to have complex dynamics and complex dependencies on membrane potential. To study this dual dependence of sodium and potassium currents on time and membrane potential, Hodgkin and Huxley applied a series of voltage steps to the squid giant axon, using the voltage-clamp technique. Their procedure was to maintain the membrane at a fixed holding potential (near the resting potential, where both sodium and potassium channels are closed) and then to step the membrane potential to a series of less negative values. They observed that such depolarizations induced an "early" negative current, followed by a "late" positive current. Remember the sign convention for membrane currents. A negative current describes the flow of positive ions into the axon. Conversely, a negative current describes the flow of positive ions out of the axon. The "early" negative current is due to ions flowing into the axon, and the "late" positive current is the result of ions flowing out of the axon. With increasingly large voltage steps, the amplitude of the negative current decreases in size. The amplitude is nearly zero for a 117 mV step (to +57 mV). For very large voltage steps there is no negative current at all; instead, an early, positive current is seen. The amplitude of the late, positive current increases monotonically as the step amplitude is increased. Thus the amplitude of both components of the ionic current are controlled by the membrane potential of the axon.

In addition to the amplitudes, the rates at which the negative and positive currents develop become increasingly large with increasingly large voltage steps. The larger rates are evident because the slopes of the negative current and positive current onsets clearly are greater with greater depolarizations. Moreover, as a consequence of increasing step size, the peak of the negative current and the switch from negative to positive current occur earlier and earlier. This means that the ionic channels open more rapidly during large depolarizations than during small ones. Hence, rates, as well as the magnitudes, of membrane currents are voltage dependent (see modeling exercise Axon-2).

I.4.5.2 Separating the total membrane current into individual ionic currents

We can explain the directions of the membrane currents observed under voltage clamp by referring to the analog model (fig. I.4-2) and equations (I.4-1 and I.4-2). First, remember that the currents in the squid axon are due to sodium and potassium ions crossing the cell membrane under voltage-clamp conditions. In other words, these ionic currents were generated with V_m, the membrane potential, set to a series of different but constant values. Also, the equilibrium potentials for the sodium and potassium ions are constant. Thus the nonconstant ionic currents in the axon must be caused by time-dependent changes in the values of the sodium and potassium conductances, g_{Na} and g_K, respectively. For relatively small voltage steps, that is, for $V_m < E_{Na}$, the electrochemical driving force for sodium ions is negative. Therefore, for small- and medium-sized voltage steps, the sodium current must be negative. On the other hand, the electrochemical driving force for potassium ions is positive for all depolarizing voltage steps, provided only that the holding potential is less negative than E_K. It is clear that the early, negative current is due to the inward flow of sodium ions and the late, positive current is caused by the outward flow of potassium ions. We can also conclude that the sodium current develops rapidly upon membrane depolarization and that the potassium current develops with some delay; hence, it is "late."

There are several plausible causes for the sign change in the membrane current. One is that the late current is larger than the early current and consequently eventually masks the inward current. A second candidate cause for the sign change is that the sodium current does not persist, that it turns off after some interval. Hodgkin and Huxley realized that to understand the shape of the current curve they needed to isolate the sodium and potassium currents. One means for finding the potassium current in isolation from the sodium current is to step the membrane potential to the equilibrium potential for sodium. Thus, when the membrane potential is stepped to about +50 mV, the electrochemical driving force on sodium ions is near zero, whereas the driving force on potassium ions is very large (about +120 mV). Therefore, during a potential step to +50 mV almost all of the current is due to the flow of potassium ions across the membrane.

To obtain measurements of isolated sodium and potassium currents for a wide range of membrane potential steps, Hodgkin and Huxley reduced the extracellular concentration of sodium ions surrounding the squid axon by replacing sodium ions in the saline solution by equal amounts of choline (large non-permeable cations). They thereby set the equilibrium potential for sodium ions, E_{Na}, to several different values. By stepping the membrane potential to these values, they obtained a family of traces for potassium currents that were due only to potassium ions. They then subtracted the potassium currents from the total currents to obtain the sodium currents (see modeling exercise Axon-3).

At this point in their experiments, Hodgkin and Huxley had described the nature of all three ionic currents: the non-time-, non-voltage-dependent leakage current and the time- and voltage-dependent early, inward current and late, outward current. Because the latter two currents were explored over a wide range of voltage steps, both the time and voltage dependence could be calculated from the voltage-clamp data.

I.4.5.3 Determining the dynamics of sodium inactivation and de-inactivation

The sodium current follows a more complex time course than the potassium current; the sodium current turns off even while the voltage step is maintained. This means that the channels that conduct the sodium ions through the membrane open, are activated when the membrane becomes depolarized, but then close (inactivate). Once inactivated, these sodium channels do not contribute to the sodium current any further until their inactivation is reversed. To fully describe the sodium inactivation processes, it is necessary to determine the rate at which the channels close following activation, to determine the voltage dependence of this inactivation, and to determine the rate at which the inactivation reverses. The reversal of inactivation, like the onset of inactivation, is voltage dependent.

Hodgkin and Huxley performed several sets of experiments to study both the rates of inactivation and of recovery from inactivation. Their approach for studying inactivation was to stimulate the squid axon with two voltage steps. The first step was a "conditioning" step for the purpose of initiating the process of inactivation. The second step was employed to test the extent to which inactivation, caused by the conditioning step, had progressed. They found that a brief conditioning depolarization leads to a reduction in the sodium current generated during the test step. The size of the current during the test voltage step becomes ever smaller as the duration of the conditioning step is increased. The cause of this reduction is that progressively more sodium channels are inactivated as the conditioning step is prolonged, leaving fewer sodium channels that can be activated when the test step is presented. From these results Hodgkin and Huxley deduced the time dependence of sodium channel inactivation.

Hodgkin and Huxley repeated these double step experiments several times to determine the rate of sodium inactivation as a function of membrane potential. These experiments included hyperpolarizing steps to remove inactivation as well as depolarizing steps to induce inactivation. The experiments demonstrated that both the magnitude and the rate at which sodium channels inactivate depend strongly on the membrane potential. For example, when the amplitude of the conditioning step was set to +29 mV, the time constant for inactivation was about 2 ms; whereas this time constant was about 7 ms for an 8 mV conditioning step. Moreover, for the larger conditioning step, unlike the smaller one, inactivation was nearly complete. That is, almost all sodium channels were inactivated by the +29 mV conditioning steps when the duration of these steps was greater than 5 ms. One interesting result was that even at the measured resting potential in the squid axon, many of the sodium channels are inactivated. These inactivated channels will not open when the axon membrane is depolarized. As we shall see later, this steady-state inactivation has important consequences for nerve impulse activity in the squid axon (see modeling exercise Axon-4).

Inactivation of sodium current is obviously not irreversible; the sodium channels progress from their closed, unopenable state that characterizes inactivation back to the openable state. In order to study the time and voltage dependence of this recovery process, Hodgkin and Huxley performed a second set of double step experiments. In these experiments they used a long conditioning step to ensure that inactivation had achieved steady-state values. Following this conditioning step, the membrane potential was set to a second constant level to allow inactivated sodium channels to recover. Finally, the axon potential was stepped to a fixed depolarized value to assess the rate of recovery from inactivation. Hodgkin and Huxley systematically varied duration of the "recovery" potential to determine the rate of the reversal of inactivation (see modeling exercise Axon-4).

The experiments to discover the rates of the inactivation processes provided the final set of empirical data on mechanisms that generate nerve impulses in the squid giant axon. Now you may ask, What do these measurements of membrane currents have to do with nerve impulses? How do they relate to membrane conductances and to the membrane potential of the squid axon under normal conditions, when it is not voltage clamped? The analytic task facing Hodgkin and Huxley, to make the connections between voltage-clamp experiments and the mechanism generating the axon impulse, was indeed formidable. As we shall see later, Hodgkin and Huxley completed the task successfully to obtain the mathematical description of membrane biophysics that still forms the core of our understanding of nerve impulses.

I.4.6 CALCULATING THE IONIC CONDUCTANCES

Recall that the aim of Hodgkin and Huxley's experiments on the squid axon was to determine the quantitative time-dependent relationship between membrane conductances for sodium and potassium ions and the membrane potential. With the data describing the time and voltage dependence of the sodium and potassium currents in hand, they could now calculate the underlying conductances. To obtain quantitative values they solved equations (I.4-1 and I.4-2) for the sodium and potassium conductances as a function of the equilibrium potentials (E_l, E_{Na}, and E_K), obtained from the Nernst equation or from direct measurements, and the membrane potentials (V_m), which were set during the voltage-clamp experiments. Hodgkin and Huxley employed these new equations to obtain conductance curves for a wide range of membrane potentials for both the sodium and potassium conductances (see modeling exercise Axon-5).

I.4.7 DESCRIPTION OF EMPIRICAL RESULTS WITH AD HOC EQUATIONS

With the extensive voltage-clamp experiments required to characterize membrane conductances completed, Hodgkin and Huxley next faced the even more daunting task of putting their results into a mathematical framework. In selecting the equations to describe their results, they received no help from molecular biology, for the structure of the cell membrane was unknown and our present understanding of the protein channels that serve as ionic conduits in the membrane was far in the future in 1952. Consequently, Hodgkin and Huxley's approach was of necessity ad hoc; that is, they chose suitable mathematical equations without much theoretical justification to fit the temporal and voltage dependence of their calculated membrane conductances for sodium and potassium ions. Their departure point for choosing an appropriate set of equations was their discernment that elementary chemical processes are described by first-order linear differential equations. Explicitly, the rate of change for a given membrane conductance g is given by

$$dg/dt = \alpha(1 - g) - \beta g, \tag{I.4-6}$$

where dg/dt is the derivative of g with respect to time t and where α and β are the forward (opening) and backward (closing) rate constants for the (then unknown) molecular events that control the conductance. Equation (I.4-6) can also be written as

$$dg/dt = (g_\infty - g)/\tau, \tag{I.4-7}$$

where

$$g_\infty = \alpha/(\alpha + \beta) \tag{I.4-8}$$

is the steady-state value of g, and where

$$\tau = 1/(\alpha + \beta) \tag{I.4-9}$$

is the time constant of the conductance change. The solution to equation (I.4-7) is

$$g = g_\infty - (g_\infty - g_0)\exp(-t/\tau), \tag{I.4-10}$$

where g_0 is the value of g at time 0. This equation describes an exponential curve with time constant τ. Depending on the specific values of the constants g_0 and g_∞, the described curve may increase or decay with time.

In comparing equation (I.4-10) to their experimental results, Hodgkin and Huxley found that it was necessary to raise g to the fourth power (g^4) to obtain a good fit between their empirical data for the potassium conductance as a function of time. To simplify their calculations they normalized their equations for the maximum membrane conductance. This allowed them to include all of the voltage and temporal dependence of the potassium conductance in a single variable n, which they named the potassium "activation." With this formulation they had

$$g_K = g_{Kmax} n^4, \tag{I.4-11}$$

where g_K is the potassium conductance, g_{Kmax} is the maximum potassium conductance (in modern terminology, the potassium conductance when all channels are open), and n is potassium activation. The variable n is described by the equations (I.4-6 to I.4-10), with n substituted for g. Conductance goes from 0 to its maximum value 1 as a function of time and membrane potential. The term n^4 represents the fraction of the total number of potassium channels that are open at any given time.

The equations that describe the activation of the sodium conductance also require that the conductance be raised to a power. To obtain a good fit to the data, Hodgkin and Huxley found that they needed to raise the activation variable m for sodium to the third, rather than to the fourth, power (m^3). Otherwise the formulation for sodium *activation* is nearly identical to that presented in equations (I.4-6 to I.4-10). The sodium conductance however *inactivates*, requiring an additional term in the formulation. Hodgkin and Huxley found that inactivation can be described by a first order equation; that is, the sodium conductance inactivates (and recovers from inactivation) as a simple exponential process. With these considerations the equation for the sodium conductance can be written as

$$g_{Na} = g_{Namax} m^3 h, \qquad \text{(I.4-12)}$$

where g_{Namax} is the maximum value of the sodium conductance (a fixed property of the membrane), m describes the sodium activation, and h describes the sodium inactivation. In this equation, m varies between 0 and 1 as the sodium conductance goes from zero to its maximum value. The term m^3 represents the fraction of the total number of sodium channels that are activated. On the other hand, h varies between 1 and 0 as the sodium channels go from fully recovered from inactivation to completely inactivated. The product, m^3h, describes the fraction of potassium channels that are open at any time and for any value of the membrane potential.

In order to provide some physical picture for these rather abstract formulations, Hodgkin and Huxley suggested that raising the potassium activation variable n to the fourth power implies that four charged particles must act in concert to open ion channels for potassium and that n is the probability that any one of these particles is in the proper position to open the channel. Similarly, they suggested that three charged particles must act simultaneously to open sodium channels and that m is the probability that any one of these particles is in the proper position to open a sodium channel. Finally, they suggested that one charged particle must act independently to inactivate sodium channels.

The complete mathematical description of the membrane potential in the space-clamped squid actions is obtained by substituting equations (I.4-11 and I.4-12) into equations (I.4-1 to I.4-3) and then substituting these new equations into equation (I.4-5). With the capacity current moved to the left side of the equation, this yields

$$C_m dV_m/dt = g_{Namax} m^3 h(V_m - E_{Na}) - g_{Kmax} n^4 (V_m - E_K) - g_l(V_m - E_l) + I. \qquad \text{(I.4-13)}$$

This equation, with the voltage dependence of all the αs and βs obtained from fitting equations to data, summarizes in succinct, mathematical form all of the voltage-clamp data obtained by Hodgkin and Huxley from their experiments on the squid axon. The complete mathematical description has come to be known as the Hodgkin-Huxley equations (see section IV.4).

I.4.8 NUMERICAL SOLUTIONS TO THE HODGKIN-HUXLEY EQUATIONS

The differential equation (I.4-13) cannot be solved analytically. That is, solutions to these equations cannot be written in closed form. Instead, Hodgkin and Huxley used numerical integration, with the calculations carried out on a calculator. Similar numerical integration is carried out today with high speed computers. Nevertheless, all of the model simulations included in this chapter are based on the formulation expressed in equation (I.4-13).

Although Hodgkin and Huxley's mathematical formulation is largely empirical, the excellent agreement between squid axon membrane potentials calculated with equation (I.4-13) and those observed in the squid axon fully vindicate their approach. In particular, the Hodgkin-Huxley equations correctly predict the shape of the nerve impulse with a high degree of precision. The primary differences are that the duration of the calculated impulse is somewhat briefer than the measured one and that the small oscillations of the after potential of the computed potential are nearly absent in the squid record. Remember, however, that the theoretical formulation, based as it is on experiments in many axons, represents averaged values. The physiological impulse was obtained from a single axon that need not reflect the average axon activity. In any case, it is clear that through their mathematical formulation Hodgkin and Huxley provided the first quantitative description of the origins of the nerve impulse. In this they achieved for dynamic neuronal membrane potentials what Bernstein had achieved for the resting potential.

Hodgkin and Huxley extended their calculations to derive theoretical descriptions of the characteristics of the traveling axon impulse. In this extension of their research, they again correctly predicted the shape and the conduction velocity of traveling nerve impulses. In particular, they correctly predicted that the depolarization associated with the rising phase of the traveling nerve impulse precedes the increase in membrane conductance. A description of the mathematical formulation is beyond the scope of this book; interested students should consult the original papers by Hodgkin and Huxley or advanced texts.

I.4.9 PROPERTIES OF NERVE IMPULSES

The formulation of Hodgkin and Huxley leads us to the following understanding of the nerve impulse. In the resting state the membrane potential is maintained at a negative potential of about -60 mV due to the leakage of some unspecified ions, which in the aggregate have a reversal potential of this value. The voltage-sensitive sodium and potassium channels are closed. (In excised squid axons, about 40% of the sodium channels are closed due to inactivation.) When the membrane is depolarized with a brief pulse of excitatory current, some sodium channels are opened. The inward flow of sodium ions through the membrane acts as a further depolarizing stimulus, leading to still further membrane depolarization. Thus the sodium current acts via positive feedback to depolarize the membrane rapidly and thereby to generate even more inward sodium current. The sodium current consequently drives the membrane potential to approach its maximum theoretical value, the sodium equilibrium potential. The peak of the impulse persists only transiently because of two restorative processes. First, the sodium channels inactivate; that is, the open channels that gave rise to the impulse close. Second, the depolarization induced by the opening of sodium channels acts, with a delay, to open the potassium channels. Because the equilibrium potential for potassium current is very negative, an outward potassium current is generated to counteract the inward sodium current. Consequently, the membrane potential becomes less positive. Rapid closing of the sodium channels (because of inactivation and reduction in membrane depolarization), combined with the open potassium channels, leads to a very rapid return of the membrane potential to the resting level. But the potential change does not stop there. The potassium channels not only open slowly, but they close relatively slowly. For some interval, the potassium channels remain open and the membrane potential undershoots the resting level to become hyperpolarized. At this potential sodium channels recover from inactivation while the potassium channels, no longer driven to an open state because the membrane is hyperpolarized, close (see modeling exercise Axon-6).

I.4.9.1 Impulse threshold and refractory period

Unless perturbed by some external signal, the giant axons of the squid remain at their resting potential. This quiescent behavior is typical of isolated axons. One means of eliciting an axonal impulse is to apply a brief excitatory current pulse of sufficient amplitude and duration. A very small pulse depolarizes the membrane slightly but does elicit an all-or-none impulse. As the amplitude of the current pulse is made progressively larger, a membrane response increments the elicited depolarizations and increases its duration. This is an active, subthreshold response that does not cause an impulse. Finally, if the pulse exceeds a certain value, an impulse is generated. This critical value is called the threshold for generating a nerve impulse. Increasing the pulse amplitude still further induces the impulse to occur with shorter latency, but it does not alter its amplitude. This well-defined threshold phenomena, found in axons of all animals, is exhibited also in impulse activity calculated with the Hodgkin-Huxley equations (see modeling exercise Axon-7).

Although the impulse threshold can be readily determined experimentally, it does not have a fixed value. For one thing, the duration of the current pulse, in addition to its amplitude, determines whether an impulse will be obtained. For very brief pulses the threshold amplitude is greater. Moreover, following one nerve impulse, the threshold to obtain a second one is greatly elevated. In fact, for a very brief interval after the peak of the nerve impulse, the threshold is so great that a second impulse cannot be elicited. This interval, the absolute refractory period, corresponds to the interval when most sodium channels are inactivated. Impulse threshold remains elevated even after the absolute refractory period because sodium channels do not all recover from inactivation at once, and because the potassium conductance is elevated, as described above, the membrane potential is hyperpolarized, at the end of the impulse. This prolonged, decreasing elevation of impulse threshold gives rise to the relative refractory period (see modeling exercise Axon-8).

I.4.9.2 Anodal break excitation

As we found above, nearly half of the sodium channels are inactivated when the squid axon membrane potential is held at the resting level of about −60 mV. If the membrane potential is sufficiently hyperpolarized with injected inhibitory (anodal) current, these channels recover from inactivation and any potassium channels that are open at this holding potential close. Abruptly releasing the axon by turning off the hyperpolarizing current can then elicit an axon impulse because the depolarizing step to the former resting potential activates sodium channel much as does a depolarizing step from the resting potential. Impulses elicited in response to a cessation of hyperpolarizing current are said to be due to anodal break excitation. Other common terms include "postinhibitory rebound" (PIR) and "paradoxical excitation." Removal of sodium channel inactivation and turning of "resting" potassium current are just two of several mechanisms that give rise to inhibition-induced excitation (see modeling exercise Axon-9).

Modeling Exercises

***NeuroDynamix* Model: AXON Exercises: Axon-1 to Axon-9**

Introduction

This set of exercises is designed to illustrate the fundamental principles underlying nerve impulse as revealed by the voltage-clamp experiments performed by A. L. Hodgkin and A. F. Huxley on the giant axon of the squid. Even after a span of more than forty years, Hodgkin and Huxley's explanation for the origin of nerve impulses in squid axon applies, with minor modification, to long-distance signaling in all nervous systems. The Hodgkin-Huxley study on the squid axon still serves as a model of scientific research at its best. View your modeling exercises as if you were carrying them out on living squid axons with Hodgkin and Huxley looking over your shoulder (fig. I.4-1). Refer to the original papers by Hodgkin and Huxley in order to compare graphs generated here by AXON with those obtained from experiments on squid axons.

Axon-1 The dynamics, amplitude, and shape of the nerve impulse

For this exercise AXON is in current-clamp mode with the stimulator set to generate current pulses of sufficient amplitude to reliably elicit an impulse each time the graph is drawn. This and all succeeding exercises in this series are based on the electrical equivalent circuit model of the squid giant axon current through the resistor (fig. I.4-2). When you begin this exercise the computer screen displays three windows: a graph of membrane potential (*Vm*), the total conductance (*gTotal*), and the equilibrium potentials for the sodium (*ENa*) and potassium (*EK*) conductances; the STIMULATOR window; and the PARAMETER window (fig. I.4-3). The rate constants are set for a temperature of 12°C.

The purpose of this exercise is to illustrate the dynamics, amplitude, duration, and shape of the nerve impulse. In addition, it demonstrates the dependence of these characteristics on the concentration of sodium in the extracellular fluid (explored by altering *ENa*).

When this exercise begins (click on the STOP/GO icon) the stimulator generates current pulses (not shown on the screen) that evoke nerve impulses at fixed intervals in the simulated squid axon. The following features are of special interest: 1) rapid upswing in potential after the initial, stimulator-induced depolarization, the fact that there is an overshoot (the potential becomes positive); 2) the amplitude of the overshoot and its very brief duration; 3) the rapid repolarization; 4) the undershoot (hyperpolarization beyond the resting membrane potential); 5) the relatively long duration of the undershoot; 6) the change of total membrane conductance during the impulse; 7) the relationship between the overshoot and *ENa*; and 8) the relationship between the undershoot and *EK*. Now lower the value of *ENa* to simulate the effect of reducing external

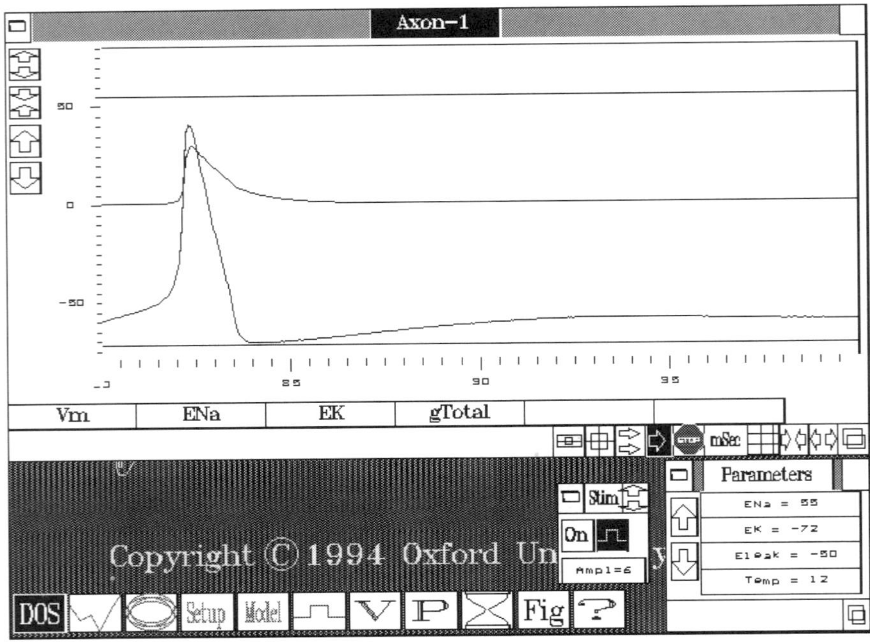

Figure I.4-3. Monitor display for exercise Axon-1

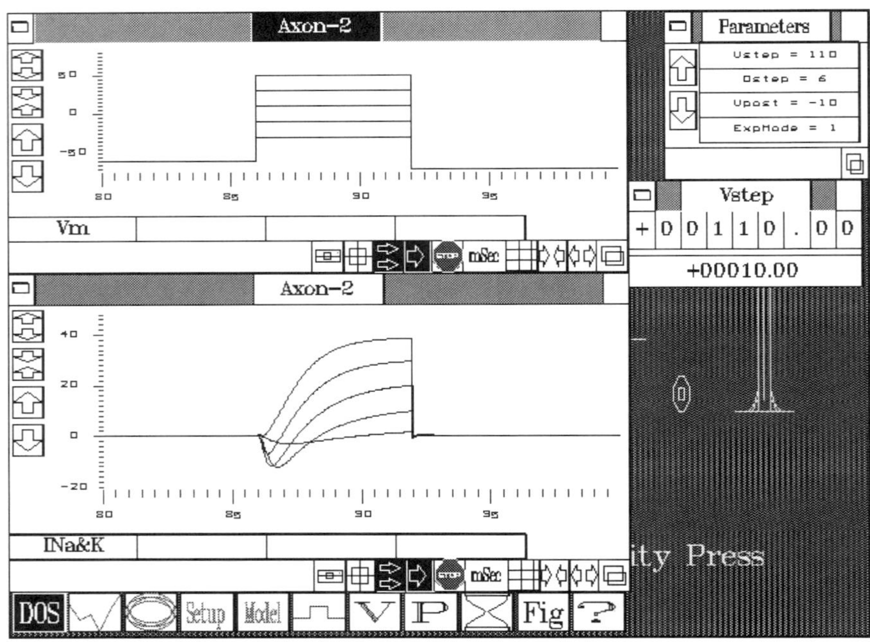

Figure I.4-4. Monitor display for exercise Axon-2

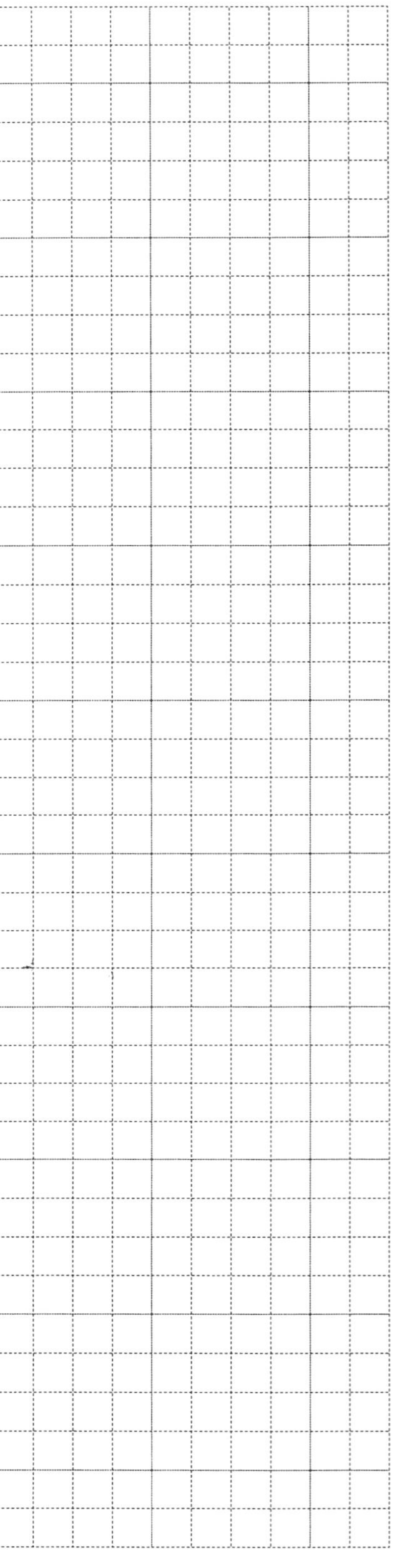

sodium ion concentration. How does this change the amplitude of the overshoot? Raise and lower the temperature. How does temperature affect the amplitude and duration of the nerve impulse? [Hint: Some of these effects are observed most effectively if traces are superimposed by clicking on the double arrows at the lower edge of the TIMESERIES window.]

Axon-2 The dynamics of membrane current in voltage clamp

For this exercise AXON is set to voltage-clamp mode in order to separate ionic currents from the capacitative current. Under voltage-clamp conditions the membrane potential is stepped to fixed values while the ionic membrane current is recorded. When you begin this exercise the computer screen displays two TIMESERIES windows, the upper one to graph the size of membrane steps (*Vm*) and the lower one to graph the time course of the summed sodium and potassium currents (fig. I.4-4). The rate constants are set for a temperature of 6°C.

To study the membrane conductance changes that generate impulses Hodgkin and Huxley depolarized or hyperpolarized the axon membrane from a fixed holding potential (*Vhold*) by a series of voltage steps (*Vstep*) of increasing amplitude. They then measured the total membrane currents before, during, and after the voltage step. The purpose of this exercise is to illustrate the dynamics of the total ionic membrane current in the squid axon with leakage current subtracted

Begin this exercise by clicking on the STOP/GO icon and observe the voltage trace. Note that the holding potential is at −60 mV. After 6 ms the potential is depolarized by 10 mV (the size of *Vstep*) for 6 ms. Finally, the membrane potential is returned to slightly below the holding potential for another 8 ms. The small depolarization of this initial voltage step does not generate much ionic current. Now open the PARAMETER MODIFICATION window for *Vstep* and increment the value of this parameter to 30 mV. Click on the STOP/GO icon to generate a new set of curves, superimposed on the previous traces. This 30 mV voltage step does cause a small membrane current that is initially negative (inward) and then becomes positive (outward). Repeat this procedure by incrementing *Vstep* by 20 mV up to a maximum value of 130 mV. This will generate a series of superimposed graphs for the membrane voltage (*Vm*) and for the total membrane current (*INa&K*; leakage current has been subtracted). Because the importance of the Hodgkin-Huxley experiments stems from their quantitative nature, measure the peak (or maximum) amplitude of the negative and positive currents for several traces (using the ANALYZE window). Then measure the delay from the beginning of the current step to the peak of the negative current. Finally, determine the time at which the current reverses (crosses the y-axis) for each voltage step. Notice that there are two prominent effects of increasing the size voltage steps on the membrane current. One is the manner in which the amplitudes of the "early" and "late" components of the current change as the voltage steps are increased; the other is the reduced time required for these components to achieve their peak or maximum value.

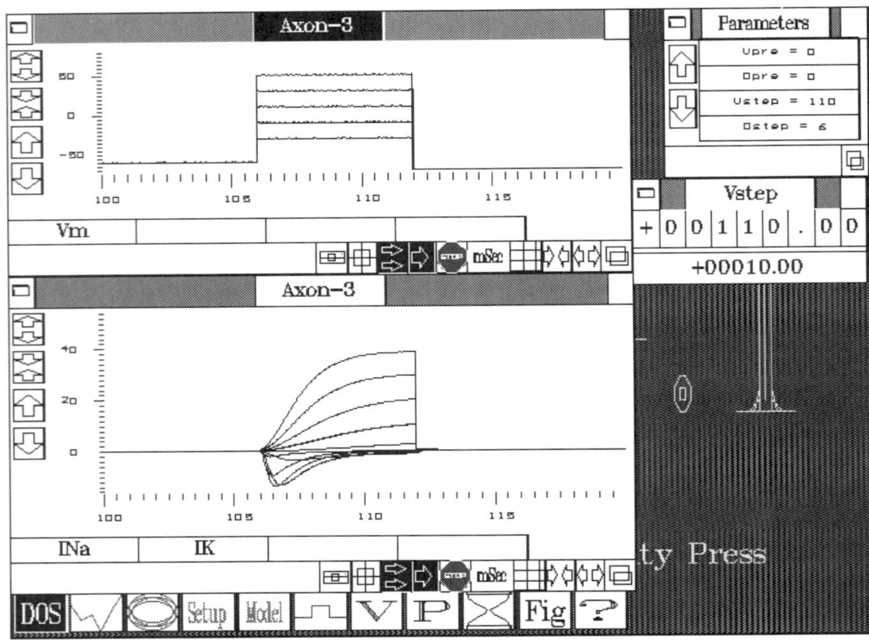

Figure I.4-5. Monitor display for exercise Axon-3

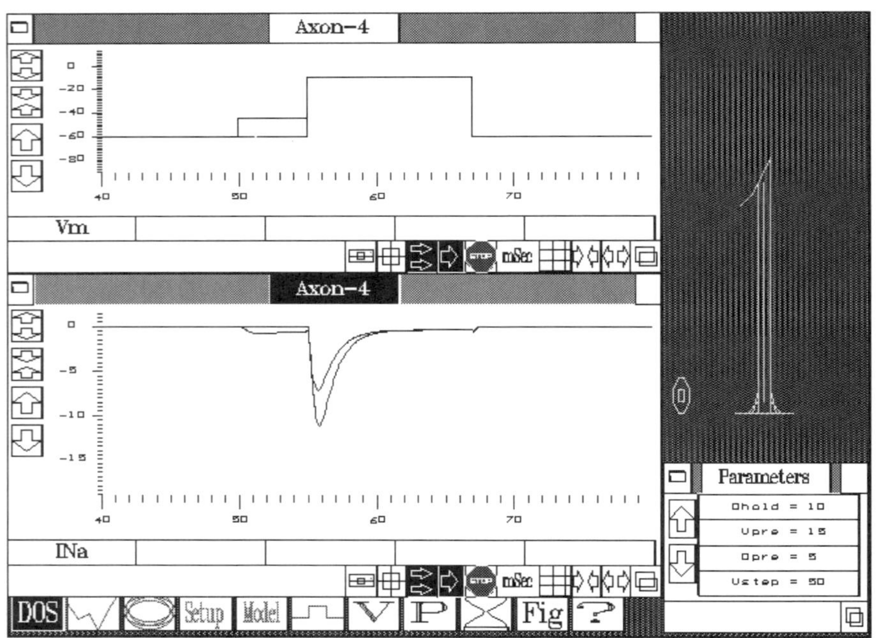

Figure I.4-6. Monitor display for exercise Axon-4

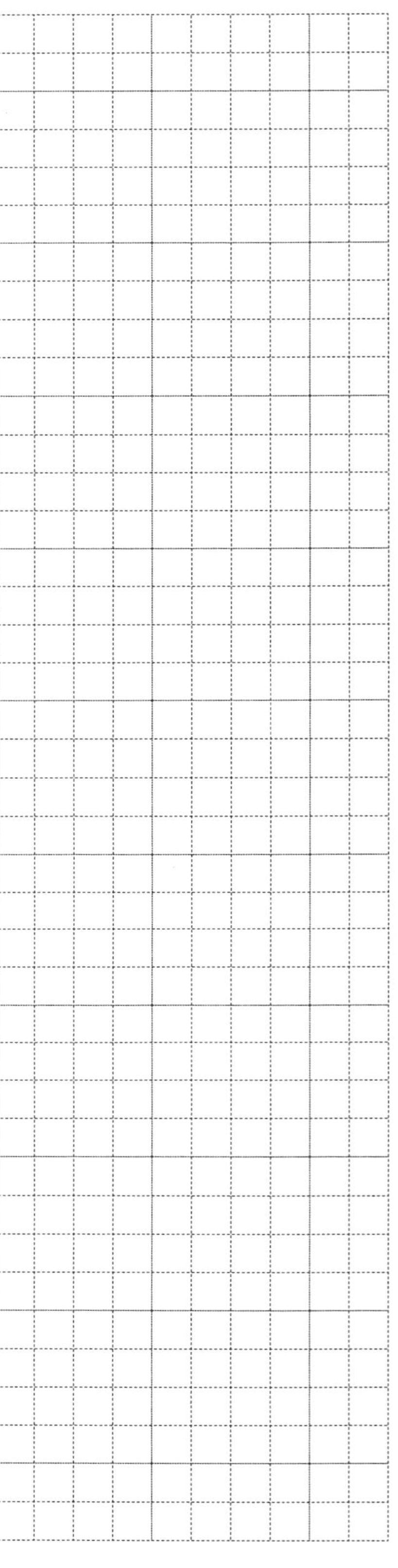

Axon-3 Dynamics of the separated sodium and potassium currents

The setup for this exercise is identical to that of Axon-2 except that the ionic current is now separated into the sodium and the currents, again viewed under voltage-clamp conditions (fig.I.4-5). This exercise explicitly demonstrates the individual dynamics of the sodium and potassium currents in response to steps in the membrane potential.

Open the PARAMETER MODIFICATION window for *Vstep* and click the STOP/GO icon to begin this exercise. Then apply small (about 10 mV) voltage steps from the holding potential in both the depolarizing and hyperpolarizing directions to observe the absence of sodium and potassium currents, *INa* and *IK*, respectively, when the potential is stepped to negative values or to small positive values. Increase the size of the depolarizing current steps to determine the smallest depolarizations that generate observable currents. Do the activation thresholds for these two currents differ? Now apply a series of increasingly larger voltage steps and observe the changes in the activation (caused by the changes in the depolarization) kinetics for both currents. Also note that the sodium current inactivates; that is, the sodium current decreases during the voltage step even though the membrane potential is constantly depolarized.

Axon-4 Dynamics of the sodium current inactivation

The setup for this exercise is similar to that of Axon-3 except that only the sodium current is graphed. The experimental procedure includes two voltage steps. The first, a conditioning step (or prepotential) that is set to a series of different durations, induces changes in the state of sodium channel inactivation. The second, or test step, is of constant amplitude and generates a sodium current to provide a measure of the degree of inactivation (or recovery from inactivation) induced by the conditioning step (fig.I.4-6). The aim of this exercise is to illustrate the rate at which sodium channels inactivate and recover from inactivation. Note that the full-scale value of time for this exercise is 40 rather than 20 ms.

At the beginning of this exercise, the amplitude of the prepotential (*Vpre*) is set to a small depolarizing value (15 mV); the duration of the holding potential (*Dhold*) is set to 15 ms; and the duration of the prepotential (*Dpre*) is set to 0 mV. When you click on the STOP/GO icon, the test potential generates a moderate-sized sodium current. Now set *Dhold* to 14 ms and *Dpre* to 1 ms to apply a brief depolarizing prepotential before the test potential. (The sum of *Dhold* and *Dpre* must be kept at 15 ms so that the sodium currents superimpose.) You will notice that the prepotential generates a very small sodium current and that the amplitude of the test current is considerably reduced. Repeat the experiment stepping *Dpre* to 2, 5, and then to 10 ms (with appropriate reductions in *Dhold*). Note that as the duration of the depolarizing prepotentials is increased, the sodium current generated by the test step gets successively smaller. Make a graph of sodium current vs *Dpre* to illustrate the rate at which the sodium channel inactivates. Repeat these procedures, but at a prepotential (*Vpre*) of −15 mv, in order to study the dynamics of recovery

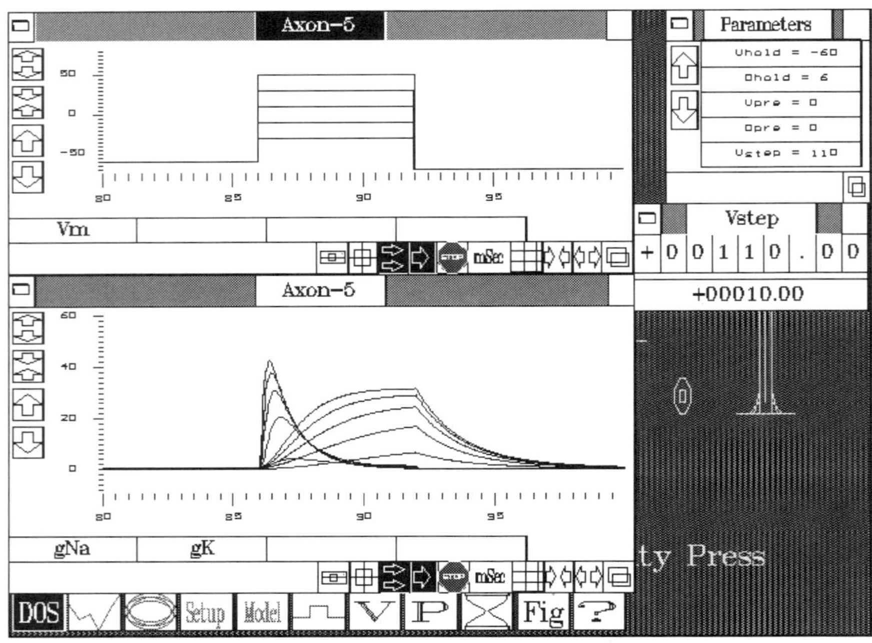

Figure I.4-7. Monitor display for exercise Axon-5

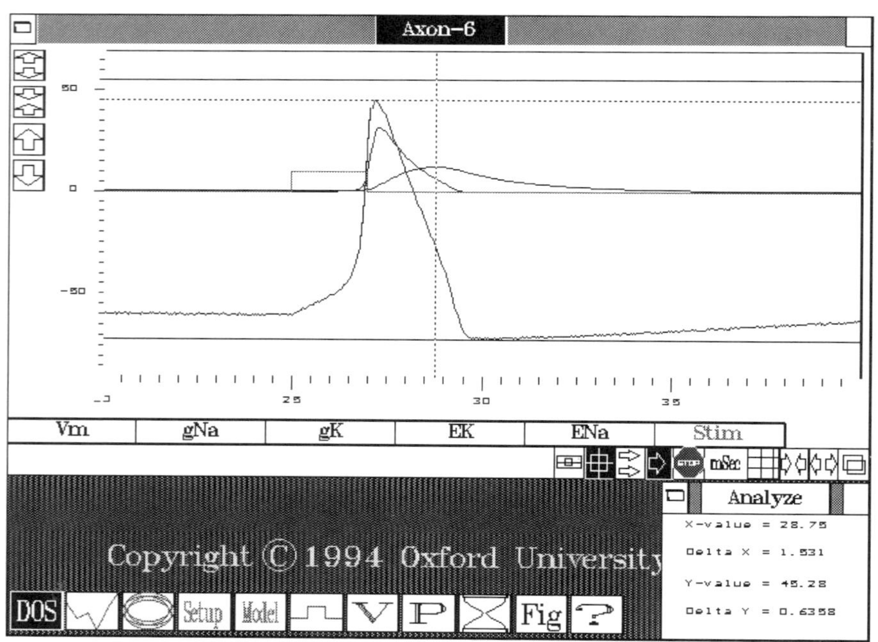

Figure I.4-8. Monitor display for exercise Axon-6

from inactivation. (Remember that the best way to erase the windows with the superimposed graphs is to open the CLOCK window and click on the CLEAR (0) icon.

Axon-5 Dynamics of sodium and potassium conductances

The setup for this exercise is identical to that of Axon-3 except that now the sodium and potassium conductances are displayed rather than the ionic currents (fig.I.4-7). Sodium and potassium conductances can be computed from the voltage-clamp experiments illustrated in Axon-3, by simply dividing the current obtained from voltage-clamp experiments by the membrane potential during the voltage step less the value of the equilibrium potential for that current. These computations are performed for you by the AXON model by solving equations (I.4-1 and I.4-2) for the conductances. This exercise illustrates the dynamics of the calculated sodium and potassium conductance values obtained in response to steps in the membrane potential.

To begin this exercise first open the PARAMETER MODIFICATION window for *Vstep* and click on the STOP/GO icon. Notice that the full-scale value of the abscissa for this exercise is 20 ms. Begin your observations of sodium and potassium conductances by applying a 10 mV depolarizing voltage step from the holding potential. Observe that this small step does not increase the sodium and potassium conductances, *gNa* and *gK*. Now apply a series of increasingly larger voltage steps (30, 50, 70, 90, 110, 130 mV) and observe the changes in the activation (caused by the changes in the depolarization) kinetics for both conductances and the deactivation (caused by repolarization) kinetics for the potassium current. Note again that the sodium conductance inactivates; that is, the sodium current decreases even as the membrane potential remains depolarized. To observe deactivation (rather than inactivation) of sodium channels, stop the graphing when the sodium conductance is at its peak value (use the STOP/GO icon). At this point set *Vstep* to zero and click the STOP/GO icon again to return the membrane potential abruptly to *Vhold*. Note that the sodium conductance deactivates very quickly, before there is any substantial inactivation.

Axon-6 Dynamics of space-clamped nerve impulses

For this exercise AXON is set to current-clamp mode as in Axon-1 with the stimulator set to generate current pulses of sufficient amplitude to reliably elicit an impulse with each stimulus. When you begin this exercise the computer screen displays only one window: a graph of membrane potential (*Vm*), sodium and potassium conductances, the equilibrium potentials for sodium and potassium currents, and the stimulus pulse (*Stim*; fig.I.4-8). For this simulation rate constants are set, like those for the voltage-clamp simulations, for a temperature of 6°C. The full-scale span of the abscissa is 20 ms. Noise is added to *Vm* to simulate the normal fluctuations observed in real neurons. The purpose of this exercise is to illustrate the relationships between the axonal impulse, the

Figure I.4-9. Monitor display for exercise Axon-7

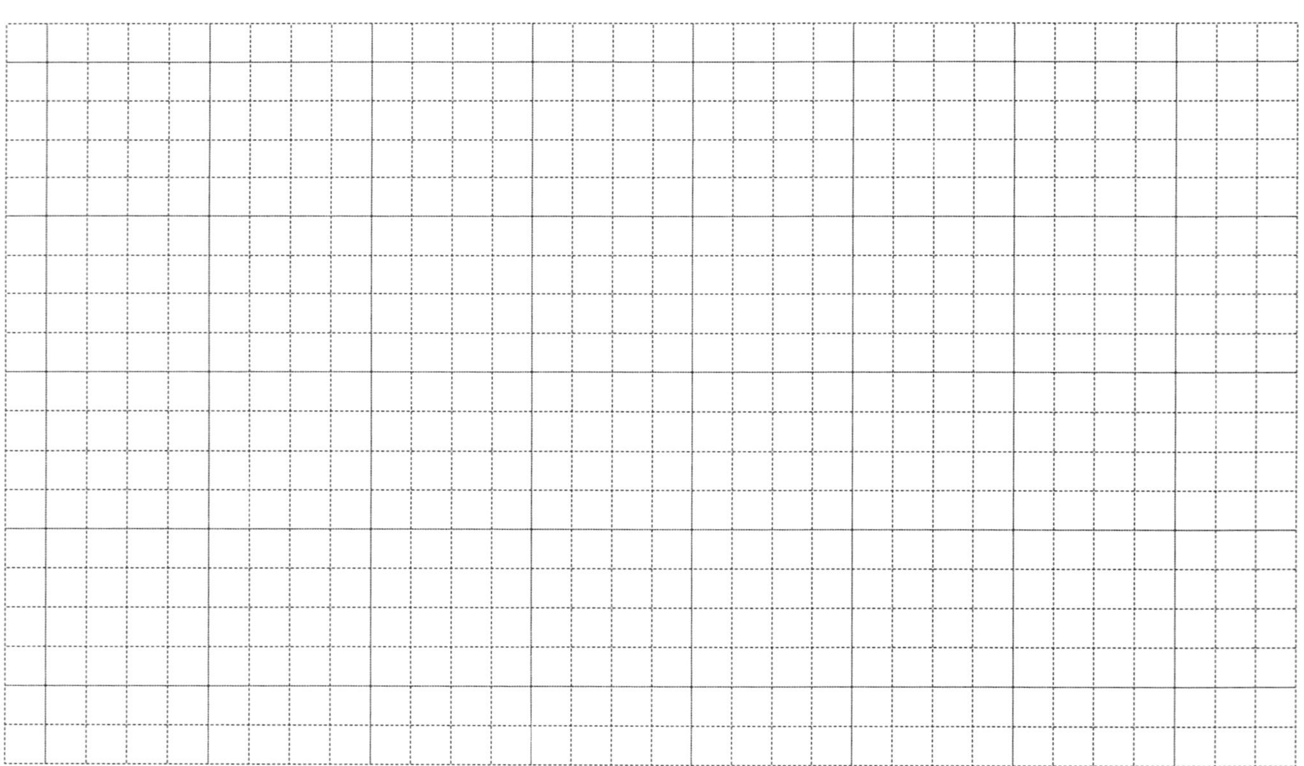

sodium and potassium conductances, and the sodium and potassium equilibrium potentials.

To begin this exercise, first open the ANALYZE window and then generate the graph twice by clicking on the STOP/GO icon once to generate the first trace and then again to generate a second one to overwrite the initializing transients seen in the first trace. You will observe that the current from the stimulator pulse depolarizes the membrane beyond threshold (about −50 mV for this exercise). Observe that the rapid rise of the impulse coincides with (and is caused by) the increase in *gNa* as threshold is exceeded. This is the positive feedback stage of the impulse, where the depolarization of the membrane causes *gNa* to increase, which leads in turn to more depolarization. Verify with the cursor that the peak of *Vm* and *gNa* occur simultaneously. What is the relationship between the value of the peak *Vm* and *ENa*? The excursion of Vm above the zero potential level is called the "overshoot." Why is the peak Vm less than *ENa*? Notice that following its brief peak, there is a rapid decay in *gNa*. Why? At this time the membrane potential begins to repolarize very rapidly, both because of the decline in *gNa* and because *gK* is finally beginning to increase. Measure the time delay between the *gK* peak and the impulse peak. Why is this peak delayed? The membrane potential hyperpolarizes past the resting potential because the potassium conductance remains high even as the sodium channels are both inactivated and deactivated. Verify that this undershoot in *Vm* is about 10 mV in this simulation to bring the membrane potential very close to *EK*.

Axon-7 Impulse threshold

For this exercise AXON is set to current-clamp mode as in Axon-6. When this simulation begins, a TIMESERIES window for graphing *Vm* and the stimulator current is open. The SIMULATOR window also is open (fig. I.4-9). The purpose of this exercise is to illustrate the threshold potential for generating nerve impulses, to show that there are active responses even when the membrane potential does not exceed threshold, and to illustrate that impulse threshold is elevated during the refractory period following an impulse. For this simulation, rate constants are again set to simulate a temperature of 6°C. The full-scale span of the abscissa is increased to 50 ms. As for Axon-6, noise is added to *Vm* to simulate the normal membrane potential fluctuations observed in real neurons.

To begin this exercise, first open the PARAMETER MODIFICATION window for the stimulator amplitude (*Ampl*); then click on the STOP/GO icon to begin graphing *Vm* as a function of time. The stimulator amplitude is initially set to a low value (1 mA/cm^2). This small current pulse induces a small depolarization each time the trace sweeps across the graph. Continue the exercise by increasing the amplitude of the current in 1 mA/cm^2 steps until threshold is exceeded. Note that this larger current pulse causes an all-or-none impulse. Now decrease the amplitude of the current pulse until impulses are no longer generated reliably. You should be able to set a value for the amplitude so that successive pulses only sometimes elicit a nerve impulse. Measure the

Figure I.4-10. Monitor display for exercise Axon-8

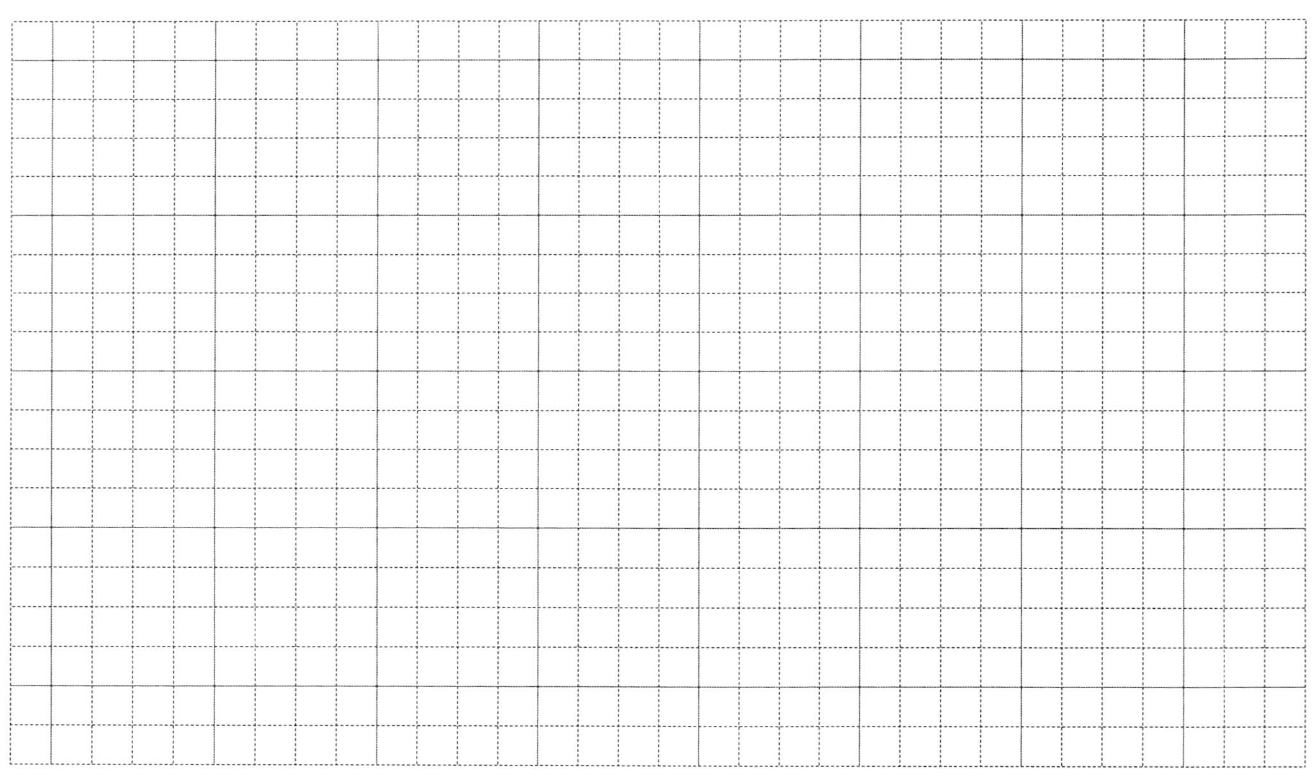

membrane potential following a pulse that does NOT generate an impulse. This is a good measure of the threshold. Note that the small depolarization caused by this subthreshold current pulse is sustained; this is an active, but still subthreshold, response. Verify that this response is caused by a small, but transient, increase in the sodium conductance by graphing *gNa* (from the VARIABLE window) at high resolution (dilate the ordinate to display +/−1.0 mA/cm^2). (Advanced students may wish to explore the phenomena of threshold further by altering the duration of the current pulse in addition to altering its amplitude.)

Axon-8 Absolute and relative refractory periods

This exercise continues the study of threshold by demonstrating the change in threshold during the refractory period that follows each nerve impulse. For this exercise AXON is set to current-clamp mode as in Axon-7. When this simulation begins, the TIMESERIES window graphs five variables: *Vm*, stimulator current (*Stim*), *gK*, *gNa*, and the percentage of sodium channels that are inactivated (*%inact*). The STIMULATOR window also is open and MAXIMIZED (fig. I.4-10). The full-scale span of the abscissa again is 50 ms. The purpose of this exercise is to illustrate that the axon is refractory following an impulse. Immediately after an impulse has occurred this refractoriness is absolute—a second impulse cannot be initiated no matter how great the stimulus amplitude. After some interval, refractoriness is relative—a second impulse can be initiated but greater stimulus amplitude is required.

To begin this exercise, first open the PARAMETER MODIFICATION window for the stimulator amplitude (*Ampl*); then click on the STOP/GO icon to begin graphing *Vm* as a function of time. The stimulator amplitude is initially set to generate individual current pulses with a value (8 mA/cm^2) that is just above threshold. After the second sweep, click the mouse on the *2x Pulse OFF* box of the stimulator window and open the PARAMETER MODIFICATION window for the *2x Pulse Delay* box. These steps reconfigure the stimulator to generate double pulses of current, with the interpulse interval set by the second PARAMETER MODIFICATION window just opened. Set the interpulse interval to 20 ms and click the STOP/GO icon to view the result, stimulating the axon with these two equal-amplitude pulses. You should observe that BOTH current pulses generate impulses. (Why?) Now reduce the interpulse interval to 10 ms. This time the second pulse does fall within the relative refractory period of the axon and, hence, does not elicit an impulse. Show that it is still possible to elicit an impulse at this delay by raising the amplitude of the current in steps until the current applied by the second pulse depolarizes the axon beyond threshold. Clearly, the threshold is relatively greater during the refractory period that follows an impulse.

Reduce the interpulse interval further and verify that a still larger current pulse is required to achieve threshold. This second impulse is now markedly smaller than the first because sodium channels inactivated during the first impulse have not had time to recover from inactivation. Fewer sodium channels are open at the peak of the second impulse; i.e., the peak value of *gNa* is

Figure I.4-11. Monitor display for exercise Axon-9

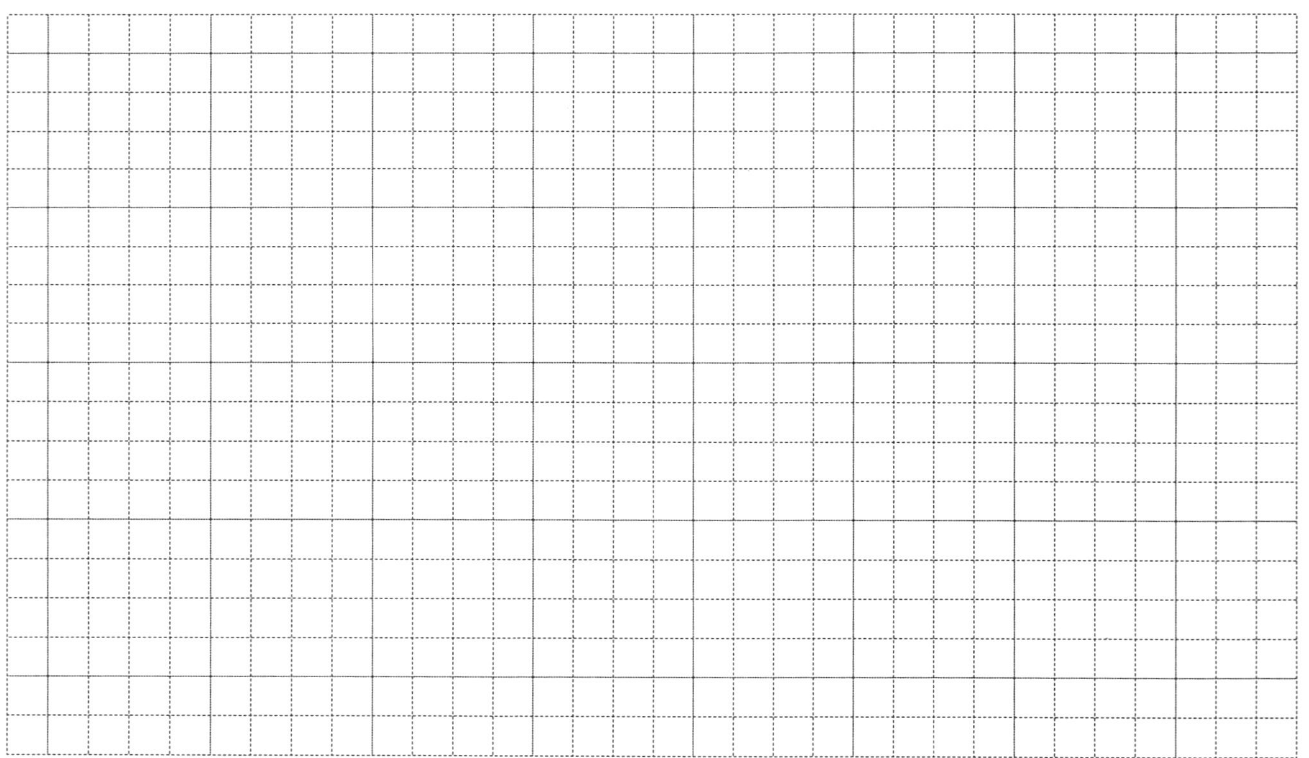

smaller. In addition, the potassium conductance has not returned to zero in time for the second impulse, and the membrane potential is still in the undershoot phase of the first impulse. Make a graph of the threshold current required to elicit an impulse versus the interpulse interval. What are the minimum and maximum durations of the relative refractory period? What is the duration of absolute refractory period?

Axon-9 Dynamics of anodal break excitation

This exercise explores anodal break excitation, a process also known as paradoxical excitation or postinhibitory rebound (PIR). For this exploration the AXON model is configured as for Axon-8, so that when simulation begins, the TIMESERIES window graphs *Vm*, stimulator current (*Stim*), *gK*, *gNa*, and the percentage of sodium channels that are inactivated (*%inact*). The STIMULATOR window also is open and is MINIMIZED (fig. I.4-11). The full-scale span of the abscissa again is 50 ms. The purpose of this exercise is to illustrate that anodal break excitation is the result of sodium-channel de-inactivation when the membrane potential of an axon is hyperpolarized by a brief current pulse.

Begin this exercise by clicking on the STOP/GO icon to begin graphing *Vm* as a function of time. Do this twice to remove the initializing transients in the first trace. Now examine the dynamics of the sodium inactivation (*%inact*) and the membrane potential (*Vm*) during and following the current pulse. Note that the current pulse (-4 mA/cm^2, *Stim* trace) briefly hyperpolarizes the membrane, which then returns to its resting level. Observe also that as the membrane potential is hyperpolarized, sodium inactivation decreases transiently. Now double the amplitude of the current pulse to -8 mA/cm^2. Notice that as a consequence of this larger current pulse, the percent inactivation (*%inact*) is greatly reduced. After the termination of the current pulse, *Vm* overshoots the resting level and then exceeds threshold to become a full-sized impulse. Note the minimum current required to generate the nerve impulse via anodal break excitation. Now double the duration of the hyperpolarizing current pulse (first MAXIMIZE the STIMULATOR window) and again note the minimum hyperpolarizing current required to elicit an impulse. Why is less current required for the longer pulse? Make a graph of amplitude versus duration for threshold stimulation that elicits current pulses by anodal break excitation. Explain the shape of this graph.

Section I.5 Properties of Neurons

I.5.1 INTRODUCTION

The neuron is an electrical machine whose inputs and outputs are chemical or electrical messages. In section I.3 we saw how membrane conductances and equilibrium potentials are the basis of the Hodgkin-Huxley-Katz model for describing the resting membrane potential. Then, in section I.4, we found that conductances and equilibrium potentials also provide the bases for understanding the nerve impulse. In this section we describe macroscopic, whole-cell conductances and explain how currents can spread between various regions of neurons—for example, from synaptic input sites to the impulse initiating region.

We have introduced two components of neurons, a passive soma and an impulse-generating axon. We now combine these with a dendrite to complete the functional neuron. Although these three compartments are characteristic of many neurons in most animals, we stress that such neuronal subdivisions are somewhat arbitrary and are important more for their heuristic value than for the fact that they are found in all neurons. A conceptual and electrical representation of the three-compartment neuron that is simulated by *NeuroDynamix* can be seen in figure I.5-1.

I.5.2 ELECTROTONIC SPREAD OF POTENTIALS WITHIN NEURONS

We know that neurons are not points in space, rather they often have very long extensions—the dendrites and the axons. For effective communication between distant regions nerve impulses are required; however for local interactions, such as between branches of dendrites and the soma, which are not mediated by nerve impulses, there is another means of electrical communication—namely, electrotonic spread of current. In the 1940s Hodgkin and Rushton realized that neurons are much like electrical cables laid on the sea floor. The important

(a)

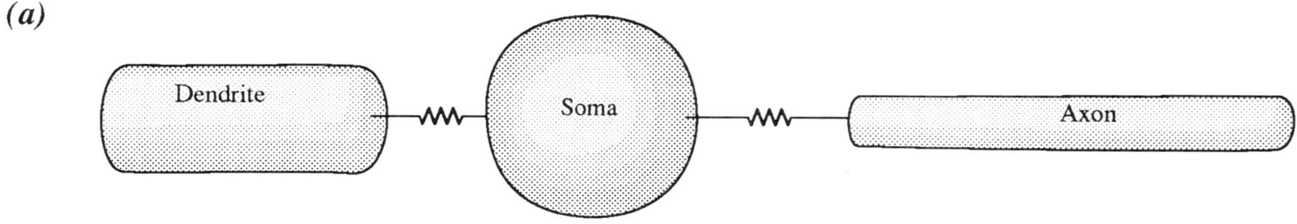

(b)

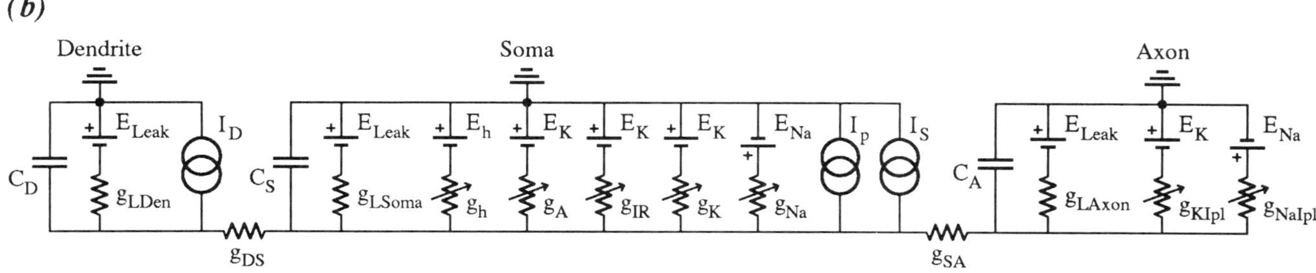

Figure I.5-1. NEURON model. *a)* Compartmental scheme for NEURON model. *b)* Circuit diagram for the three NEURON compartments.

characteristics of such marine cables are the resistance of the central wire conductor, the insulating properties of the surrounding sheath, and the action of sea water as an electrical conductor. Note that neurons similarly include an internal conductor—the cytoplasm, an insulating sheath—the cell membrane and that these too are surrounded by a conducting fluid—the extracellular milieu. Hodgkin and Rushton adapted the physical theory developed by Kelvin and Heaviside in the late nineteenth century for marine cables to electrotonic communication in neurons, much as Bernstein adapted Nernst's theory to describe cell resting potential.

The details of cable theory are outside the scope of this book. Briefly, in neurons as in marine cables current flows downhill to regions where the potential is more negative (or less positive). Similarly, if one region of a neuron is depolarized (positive) with respect to another, there will be current from the depolarized to the more polarized region. This current then acts to depolarize this nearby region. The extent of this spread of a depolarized (or hyperpolarized) potential is characterized by a number, the length constant λ (lambda). For neurons with simple geometries, say for a cylindrical dendrite, the electrotonic spread of a locally generated potential is given by the equation

$$V_x = V_o * e^{-x/(\lambda/n)} , \qquad \text{(I.5-1)}$$

where V_o is the potential at some local point and V_x is the potential at some distance (x) from this point. For these simple geometries, equation (I.5-1) states that the membrane potential decreases exponentially with distance. The rate of this decrease is given by λ/n. For structures with a large value of λ, the local changes in potential spread a long distance, whereas a small value for λ means that the potential change does not spread far. Recall that in exponential equations such as (I.5-1), V_x will be about 1/3 (actually 0.37) of V_o when x equals λ.

For the simple, cylindrical dendrite or axon, the value of λ is

$$\lambda = (0.5 * a * R_m/R_i)^{1/2}, \qquad \text{(I.5-2)}$$

where a is the radius of the cylinder, R_m is the specific membrane resistance, and R_i is the specific resistance of the cytoplasm. (An implicit assumption is that the resistance of the extracellular milieu is negligible.) Note that for a fixed, given set of resistance values, the extent to which potentials spread electrotonically increases with the square root of the axon or dendrite diameter.

The concept of length constant is useful even when it cannot be easily calculated because of complex neuronal geometries. The critical point to remember is that currents, and therefore potentials, spread within cells without the need for nerve impulses. The extent of the spread is indicated by the length constant. Neurons, or neuronal components that are characterized by large length constants have nearly the same potential throughout. Conversely, neuronal regions with small length constants will only act locally in the absence of nerve impulses. In general, small neuronal processes, with low membrane resistance (high conductance) will have a small length constant, whereas in large-diameter structures, with high membrane resistance, the length constant will be large.

The foregoing description applies to steady-state potentials. For brief or rapidly changing potentials, there is another consideration, namely, the time constant of the neuronal membrane. In considering electrotonic conduction it is useful to think of neurons as low-pass filters (i.e., filters that allow low-frequency signals to pass but that attenuate high-frequency signals strongly). What we have described is the direct current (DC), non-changing situation. Signals that change with time, such as nerve impulses or synaptic potentials are more strongly attenuated by electrotonic interactions than the DC ones. Thus, a slow synaptic potential in a dendrite may spread with little attenuation to the axon hillock, but an impulse generated at the hillock may cause hardly a ripple in the dendritic membrane potential (see modeling exercise Neuron-1).

I.5.3 REFRACTORY PERIOD AND REPETITIVE FIRING

Neurons are capable of generating repetitive impulses whose frequency is governed by properties intrinsic to individual neurons and by the intensity of the input. We describe here some of the factors that determine the rate at which impulses arise in a neuron stimulated by a constant, excitatory input.

One important neuronal property that sets the interval between impulses is the refractory period. Following each impulse there is an interval called the "refractory period" during which the neuron is less excitable (section I.4). During this period a greater-than-normal stimulus is required to elicit another impulse. Immediately following each impulse there is an *absolute* refractory period, during which even a huge stimulus is ineffective. The later refractory period is named the *relative* refractory period because relatively large stimuli can now elicit a nerve impulse. The degree of refractoriness declines following each impulse; hence, nerve impulses are generated during continuous membrane depolarization each time the refractoriness (or, in other words, when the threshold for impulse initiation) has declined sufficiently so that the stimulus is above threshold for initiating an impulse once again.

There are several cellular properties that contribute to the refractory period. One of these is sodium channel inactivation. Recall from experiments on AXON (section I.4) that the membrane potential repolarizes following the impulse peak because sodium channels inactivate (convert from the open to the closed–not-openable state). Immediately following an impulse most sodium channels are inactivated and therefore not available for generating another impulse. Sodium inactivation is reversed gradually during the hyperpolarizing afterpotential following an impulse. This removal of inactivation is one process that serves to end first the absolute, and then the relative refractory, period.

Following the nerve impulse peak, potassium channels open, leading to membrane potential repolarization. Because potassium currents counteract excitatory inputs, the opening of potassium channels (increased g_K) is a second neuronal property that contributes to the relative refractory period. During the relative refractory period the potassium channels gradually close. Closing of these channels together with recovery from inactivation of sodium channels end the relative refractory period; that is, the threshold for impulse initiation is reduced by these processes to permit the initiation of another impulse.

Alterations in the states of sodium and potassium channels are most important for determining the impulse frequency of nerve impulses (that is, in setting the intervals between impulses) when impulses occur in rapid succession. Cells can also generate impulses at very low frequencies, either due to discrete synaptic inputs or as the result of a pacemaker potential. This latter potential is a slow depolarization, caused by a small inward current, which gradually brings the membrane potential above threshold for generating an impulse. The size of this current determines the rate at which impulses repeat in slowly spiking cells. As an example, if the leakage conductance is increased in AXON model simulations of the current-clamped squid axon, the simulated axon will generate a continuous train of nerve impulses (see modeling exercise Neuron-2).

I.5.4 VOLTAGE SENSITIVITY OF ION CHANNELS

For the voltage-activated conductances simulated by *NeuroDynamix,* the relationship between the macroscopic channel conductance and the membrane potential is sigmoidal. The conductances vary from 0 (when all channels are closed) to some maximum value, g_{max} (when all channels are open), as the membrane potential becomes less negative (more negative for two conductances described later, g_{IR} and g_h). This sigmoidal relationship is described by the following equation:

$$g = g_{max} * \{1 + \exp[-(V_m - V_{hlf})/V_{slp}]\}^{-1}, \qquad (I.5\text{-}3)$$

where g_{max} is the conductance when all channels are open, V_{hlf} is the membrane potential at which half of channels are open (or individual channels are open half of the time), and V_s is the voltage sensitivity of the channel. The inverse of this last term is a measure of the number of charges moving through the membrane when a channel is gated open. For conductances that are activated by hyperpolarization, the V_{slp} term is negative.

I.5.5 AN INTRODUCTION TO FOUR CHANNELS AND CURRENTS

This is a very brief introduction to the currents generated by three types of potassium channels and by an odd, mixed sodium-potassium channel. Each of these channel types may have several related subtypes.

I.5.5.1 Delayed rectifier (I_K)

This voltage-gated potassium channel is opened by depolarization. Because it has very slow inactivation kinetics, it closes only slowly during prolonged stimulation. One of the functions of the K channel, as we saw from the Hodgkin-Huxley experiments on the squid axon, is to repolarize the membrane after each impulse. It is identified pharmacologically by application of the blocker TEA (tetraethylammonium). The reversal potential for the current through this channel is at the potassium equilibrium potential, E_K.

I.5.5.2 Fast transient current (I_A)

This second channel also is a voltage-gated potassium channel opened by depolarization; however, less depolarization is required to open this channel than to open the delayed rectifier channel. Unlike the delayed rectifier, this channel has fast inactivation kinetics. These channels open only briefly (tens of ms) and then pass into a closed, unopenable state much like that of voltage-gated sodium channels. Like in sodium channels, inactivation of the fast transient channels is removed by hyperpolarization. One effect of I_A is to delay the onset of impulse activity at the onset of an excitatory stimulus; their presence also acts to reduce the overall impulse frequency in stimulated cells. The fast transient channels are blocked by 4-AP (4-aminopyridine). The reversal potential for the current through this channel again is at E_K (see modeling exercise Neuron-3).

I.5.5.3 Inward rectifier, anomalous rectifier (I_{IR})

Unlike I_K and I_A, this potassium current is activated when the membrane potential is hyperpolarized below, rather than depolarized above, the resting level. Because these channels are closed for potentials less negative than E_K, currents through the inward rectifier channels do not reverse; they are inward (depolarizing) whenever the membrane potential is sufficiently negative to gate them open. The presence of I_{IR} is detected by the concave upward curvature of *I-V* curves for large, hyperpolarizing currents. In heart muscle, this current serves to reduce membrane conductance during the depolarized sector of the cardiac cycle. The inward rectifier current is blocked by application of a small membrane depolarization.

I.5.5.4 Sag current (I_h)

Like I_{IR}, this current is activated when the membrane potential is hyperpolarized; the channels activate when the membrane potential goes more negative than about −50 mV. Unlike the three other channels described here, the current through this channel, I_h, is the result of ion flow generated by the movements of both sodium and potassium ions. Because of the mixed nature of this current, the reversal potential is about −20 mV, about halfway between E_{Na} and E_K. When the membrane potential is hyperpolarized with a current pulse, the I_h channels open, generating a slow inward current that counteracts the hyperpolarization. This depolarization leads to a sag in the membrane potential, hence the origin of the name "sag" current for this hyperpolarization-activated current. When the hyperpolarizing current is terminated, the membrane potential overshoots its resting level because, with its rather slow kinetics, I_h remains activated for several seconds, depolarizing the membrane. In cells that include this conductance, hyperpolarizing pulses are followed by postinhibitory rebound (PIR). Conversely, depolarizing current pulses reduce the activation of I_h. Therefore, there is a negative, hyperpolarized afterpotential following depolarizing pulses. The recovery from this transient hyperpolarization coincides with the partial reactivation of the inward current. The I_h current is blocked by extracellular application of cesium ions (see modeling exercise Neuron-4).

I.5.6 STEADY-STATE *I-V* CURVES

A useful picture for neurons, as we saw earlier, is the parallel conductance model, in which the cell membrane is viewed as a parallel arrangement of resistors and batteries, additional current paths, and a capacitor. One approach to assessing the value of the resistors in this model is via Ohm's law. Physiologists determine the value of membrane resistance by passing a measured current into a cell while simultaneously monitoring the membrane potential. For simple membranes (constant resistance value) Ohm's law states that the resistance is simply the change in the membrane potential divided by the amplitude of the injected current (section I.3). Cell membranes, however, are seldom simple; the resistance of the cell membrane usually changes as the amplitude of the current is increased. To assess this change in resistance, a series of current pulses are injected into a given neuron and the membrane potential excursions are graphed as a function of the current amplitude. This experimental procedure generates in a voltage versus current (*I-V*) graph whose slope at any current is the membrane resistance. (The reciprocal of the slope is the membrane conductance.)

The slope of the *I-V* graph is a straight line only if the cell membrane includes no voltage-dependent conductances. Curved *I-V* graphs imply that the membrane acts as a rectifier. For most neurons the *I-V* graph is sigmoidal; that is, the curve flattens for both large positive and large negative currents. Flattening for positive currents results from the depolarization-induced opening of potassium channels (such as those that terminate the nerve impulse). Flattening of the curve for hyperpolarization is caused by the opening of another class of potassium channels, inward rectifier channels (g_{IR}) that are opened by hyperpolarization (see modeling exercise Neuron-5).

Modeling Exercises

***NeuroDynamix* Model: NEURON Exercises: Neuron-1 to Neuron-5**

Introduction

This series of modeling exercises is designed to help you understand the physiology of real, spatially non-uniform neurons. For these exercises we are using the three-compartment NEURON model, which simulates a neuron that comprises a dendrite, a soma, and an axon (fig. I.5.1). These compartments are linked via electrical coupling whose strength is under experimental control. The experimenter controls the function of this model cell through currents injected into the dendrite (to simulated synaptic input) and into the soma. NEURON simulates a variety of currents, including three potassium currents, a slow, inactivating sodium current, a hyperpolarization-activated inward current, and the action of an electrogenic pump. Only a few of the many functions of NEURON are illustrated in the exercises described below; you may wish to do additional exercises on your own.

Neuron-1 Electrotonic interactions between neuronal compartments

For this exercise NEURON is configured to simulate a linear, three-compartment neuron with a dendrite, a soma, and an axon. Each of these three compartments has a resting conductance and a capacitance, which together determine the membrane time constant. The axon includes also fast, active sodium and potassium conductances which simulate (but much more crudely than in AXON) nerve impulses. The soma compartment includes several conductances that are both voltage- and time-dependent (see fig. I.5-1b). When you begin this exercise, the computer screen displays four windows: 1) a TIMESERIES that displays the dendrite (*VmD*) and soma (*VmS*) membrane potentials; 2) a second TIMESERIES window that displays the membrane potential in the axon compartment (*VmA*, note the difference in voltage scales); the PARAMETER window; and the STIMULATOR window (fig. I.5-2). The aim of this exercise is to illustrate the relationships between membrane potentials observed in the three compartments of a spatially distributed neuron.

When you begin the exercise (click the STOP/GO icon), the stimulator is connected to the dendrite (parameter *StimD* = 1) and not to the soma (*StimS* = 0). With the stimulator current set to zero, all three neuronal compartments are at the resting potential of −70 mV; because the threshold in the axon is set to −69 mV, no impulses are generated. Now pass a 1 nA current into the dendrite; set *Ampl* on the stimulator to 1 nA, simulating a constant excitatory synaptic input. Note the three different potential excursions caused by this stimulus in the three compartments. The largest depolarization (about 15 mV) occurs at the site of current injection, the dendrite (*VmD*, upper window); a smaller depolarization (about 4 mV) is seen in the soma (*VmS*, upper window);

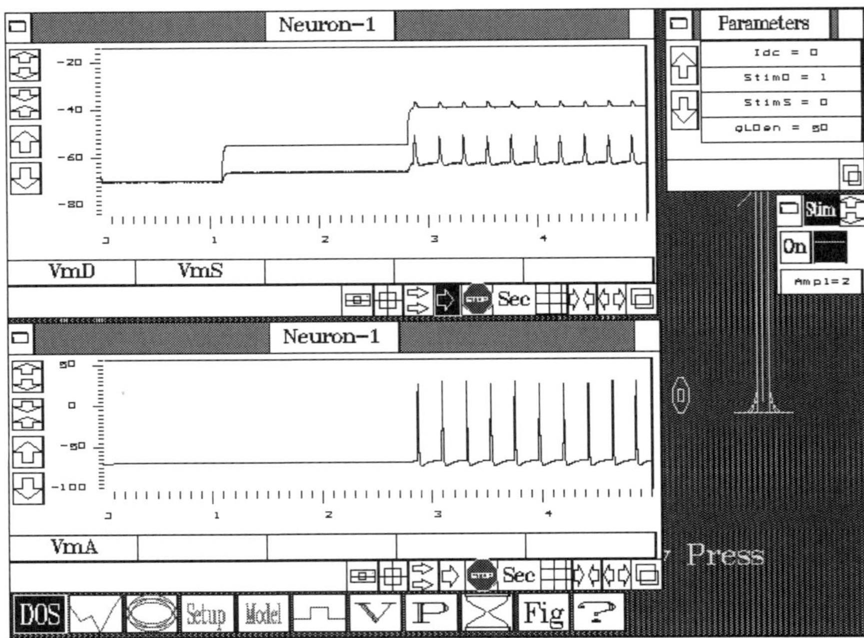

Figure I.5-2. Monitor display for exercise Neuron-1

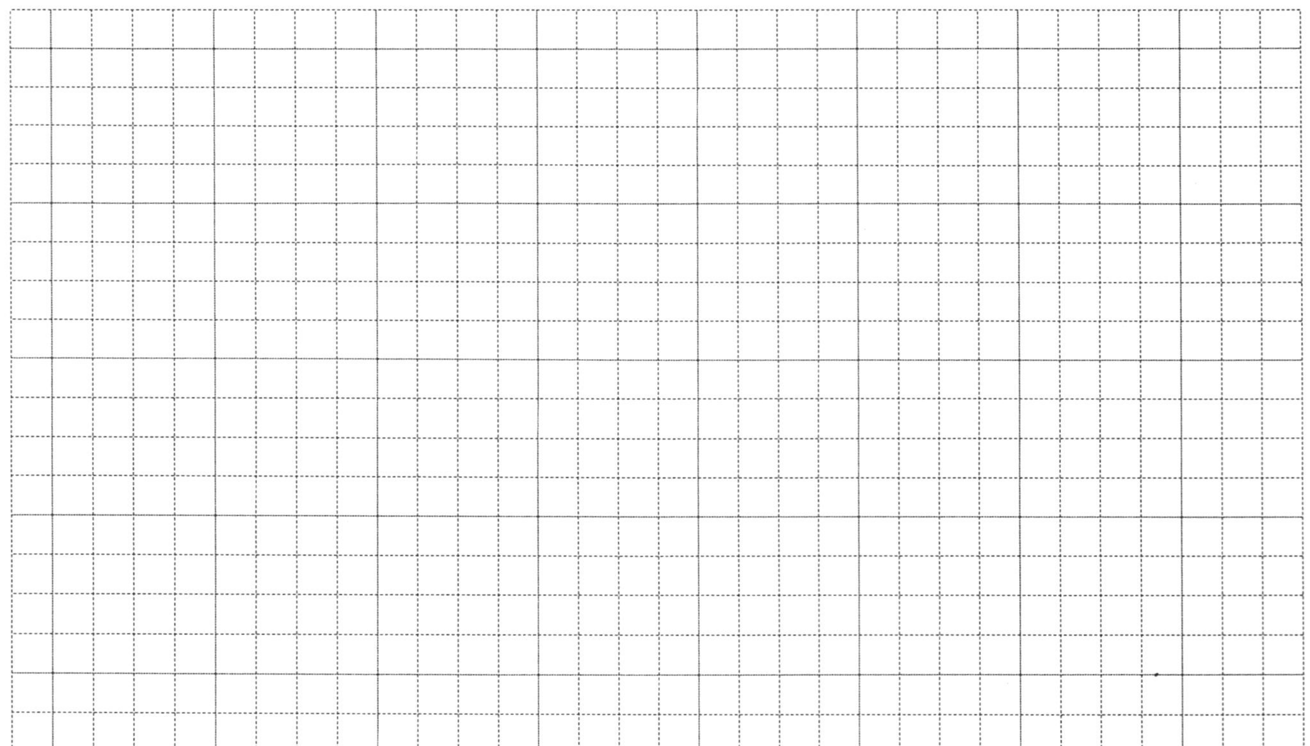

and only a very small depolarization (about 1 mV) reaches the axon (*VmA*, lower window). No impulses are generated by the 1 nA current because the small depolarization in the axon is insufficient to exceed threshold. We can quantify the attenuation between the neuronal compartments by noting that the dendrite-to-soma ratio for the depolarizations is about 4:1; similarly, the soma to axon ratio also is roughly 4:1.

Increase the stimulus amplitude to 2 nA so that the axonal membrane potential exceeds threshold. In the NEURON model fast sodium and potassium conductances are found only in the axon compartment; hence the overshooting, 100 mV impulses are seen only on the *VmA* trace in the lower window. These full-sized impulses are reflected in the soma as small, sharp depolarizations and as very small bumps in the dendrite. By measuring and comparing the amplitude of these impulse-generated deflections in the three compartments, you will discover that impulse attenuation is considerably greater (nearly 10:1) than for the depolarizing steps. This comparison illustrates the general result that high frequency signals undergo much greater spatial attenuation than low-frequency, d.c. (direct current) signals.

Reduce the amplitude of the stimulator current to 1 nA in order to stop impulse activity. Then set the stimulator to pass this current into the soma rather than into the dendrite (set *StimD* = 0 and *StimS* = 1). Notice that although 1 nA is below impulse threshold when passed into the dendrite, it does exceed threshold when passed into the soma. This simple demonstration illustrates why synaptic connections to neuronal somata are more effective for controlling impulse activity than synaptic connections to dendrites.

Finally, you can make some more detailed measurements on the coupling of potentials in the three compartments of this model. First, turn the stimulator off by setting the amplitude to 0. This returns all compartments to the resting potential. Expand the ordinates on both graphs and open the ANALYZE windows to make accurate measurements of changes in membrane potential. Now apply −1 nA to the soma and measure the changes in membrane potentials in the soma, dendrite, and axon compartments induced by this negative current. Calculate the dendrite-to-soma ratio and the axon-to-soma ratio of the attenuations. These ratios provide a good measure of the electrotonic coupling between the three compartments. The larger the ratio, the tighter the coupling *and* the less the attenuation of electrical signals passing from one compartment to the other. Repeat this procedure and the calculations for −1 nA injected into the dendrite. Note also that the membrane voltage changes when currents are applied to the dendrite, are most rapid in the dendrite, and are slowest in the axon compartments.

Neuron-2 Control of impulse frequency

For this exercise NEURON is reconfigured to simulate a more simple, two-compartment neuron that consists only of a soma and an axon (fig. I.5-3a). These compartments have the same properties as those of Neuron-1; however, the coupling between compartments is closer. Because of this tighter coupling, impulses generated in the axon are seen in soma records as larger deflections.

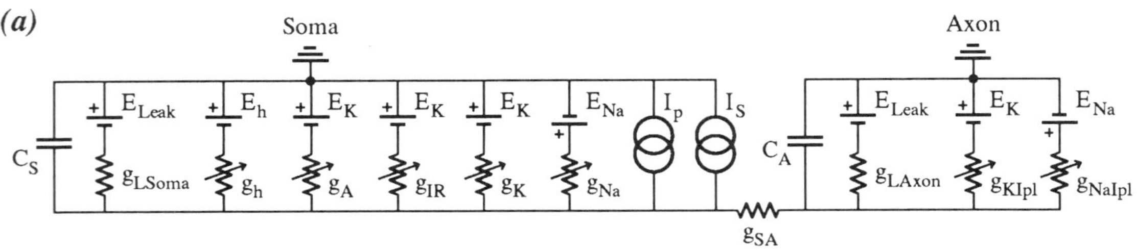

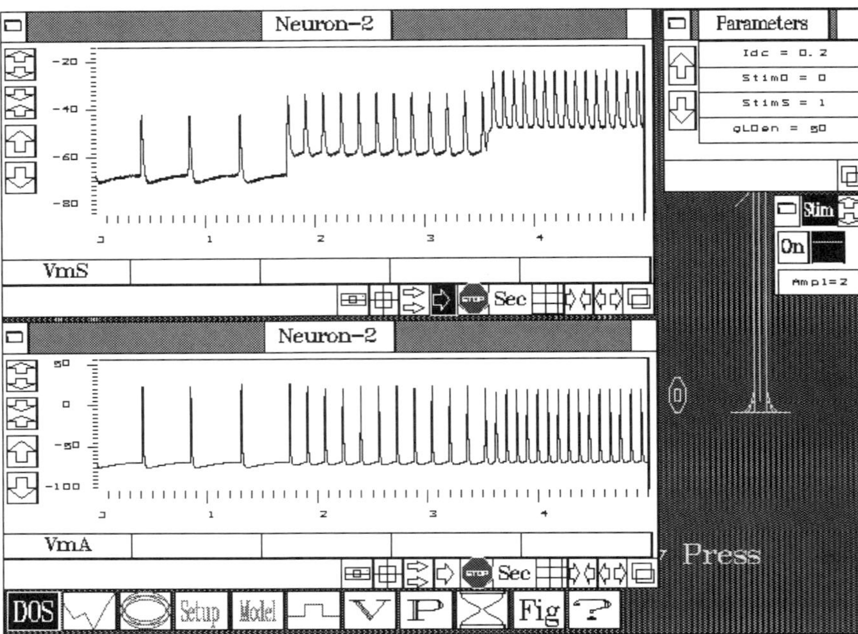

Figure I.5-3. Two-compartment NEURON model. *a*) Circuit diagram of the reduced two-compartment model. *b*) Monitor display for exercise Neuron-2

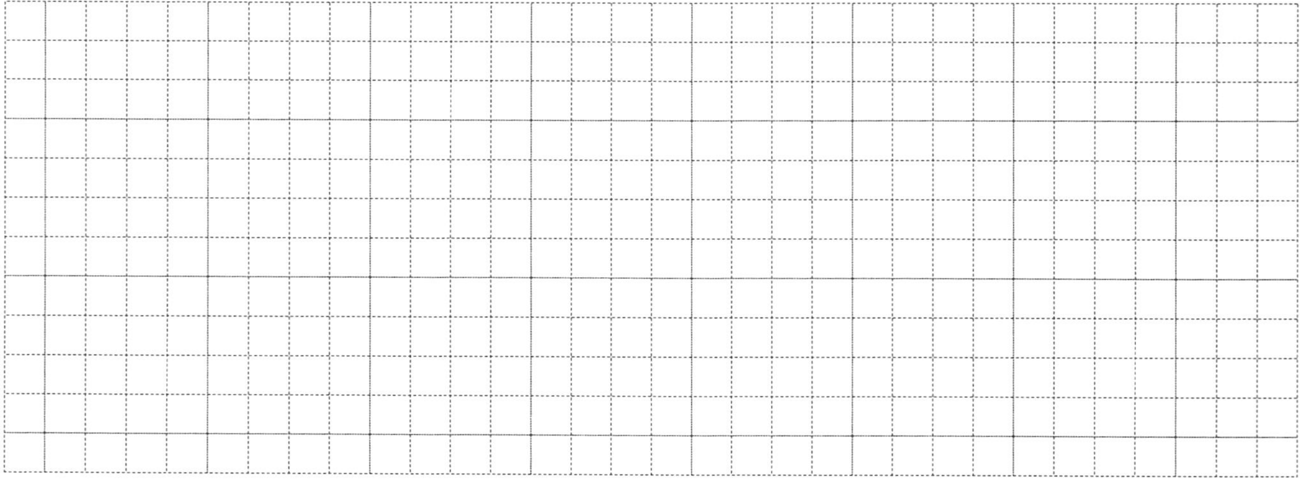

Four windows are open for this exercise: a TIMESERIES window that displays the soma membrane potential (*VmS*); a second TIMESERIES Window that displays the axon membrane potential (*VmA*); the PARAMETER window; and the STIMULATOR window (fig. I.5-3b). The aim of this exercise is to show how the rate at which impulses are generated by the axon depends on the amplitude of current injected into the soma.

At the beginning of this exercise, a small depolarizing current is injected into the soma to bring the axon membrane potential above threshold (*Idc* = 0.2 nA). At this low level of excitation the impulses occur at intervals of about 0.5 s; that is, the firing rate is about 2 Hertz (Hz). Increase the firing rate by setting the stimulator to 1.0 nA. (The current is directed to the soma compartment.) Use the ANALYZE window for the lower graph to verify that the inter-impulse interval is about 0.16 s; hence the frequency is about 6 Hz. Increment the stimulus current by 1 nA steps, up to 5 nA, and make a graph of the impulse frequency as a function of current. This graph is nearly linear for small currents but saturates (at about 30 Hz) for currents above 3 nA. When the frequency-vs-current curve is extrapolated to zero impulses, the threshold current (where the extrapolated line crosses the x-axis) should be near 0 nA. Determine the precise value for the threshold current by setting the stimulator amplitude to zero and varying *Idc* until impulses occur at intervals greater than 1 s. [Hint, open the PARAMETER MODIFICATION window for *Idc* and adjust current levels in 0.01 nA steps.] Note that when the simulated neuron is firing at this low rate, the inter-impulse interval is variable. This variability is due to the noise superimposed on the soma membrane potential.

Neuron-3 Role of the A current

For this exercise NEURON is configured to simulate the simpler, two-compartment neuron with a soma and an axon that was used in Neuron-2 (fig. I.5-3a). Resting parameters of these compartments are as in Neuron-2. For this exercise, the soma includes an A current, which is a time-dependent potassium current that activates and then inactivates upon depolarization. When you begin this exercise the computer screen displays four windows: 1) a TIMESERIES window that displays the soma (*VmS*) membrane potential; 2) a second TIMESERIES window that displays the A conductance (*gA*) and the axon potential (*VmA*); the PARAMETER window; and the STIMULATOR window (fig. I.5-4).

The aim of this exercise is to illustrate the role of the A conductance in regulating the rate of impulse frequency in neurons. Begin simulation by clicking on the STOP/GO icon; at this time the A current is turned off (the A channels are blocked). As the traces move across the screen, apply 1 nA of current to the soma by turning the stimulator on. Notice that the soma depolarizes quickly and immediately exhibits a series of impulses (remember that these are the electronically conducted signals from the axon) that follow each other at a constant rate. Turn the stimulator off and open the PARAMETER MODIFICATION window for the A conductance (*gAMax*) and set its maximum value to 400 nS. The membrane potential will shift to a slightly more

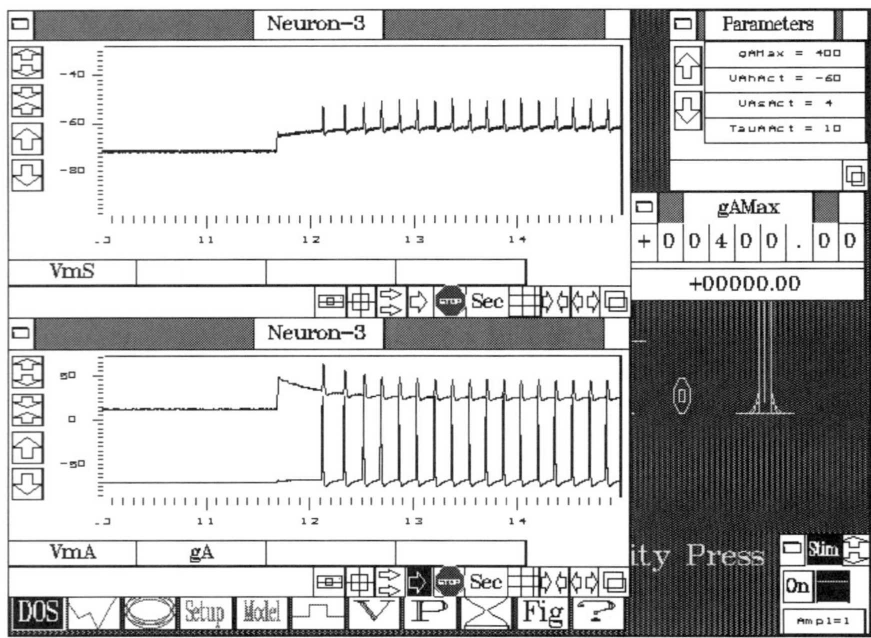

Figure I.5-4. Monitor display for exercise Neuron-3

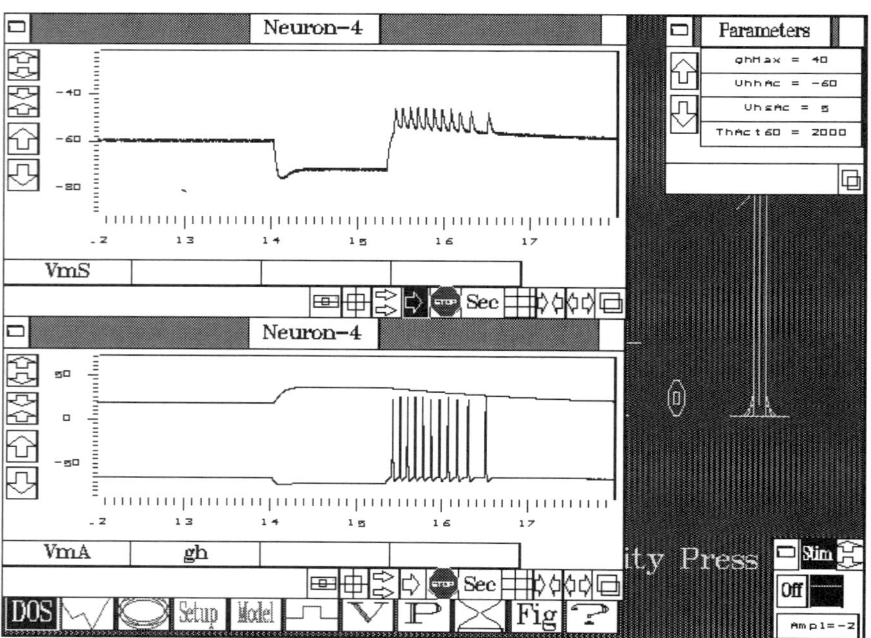

Figure I.5-5. Monitor display for exercise Neuron-4

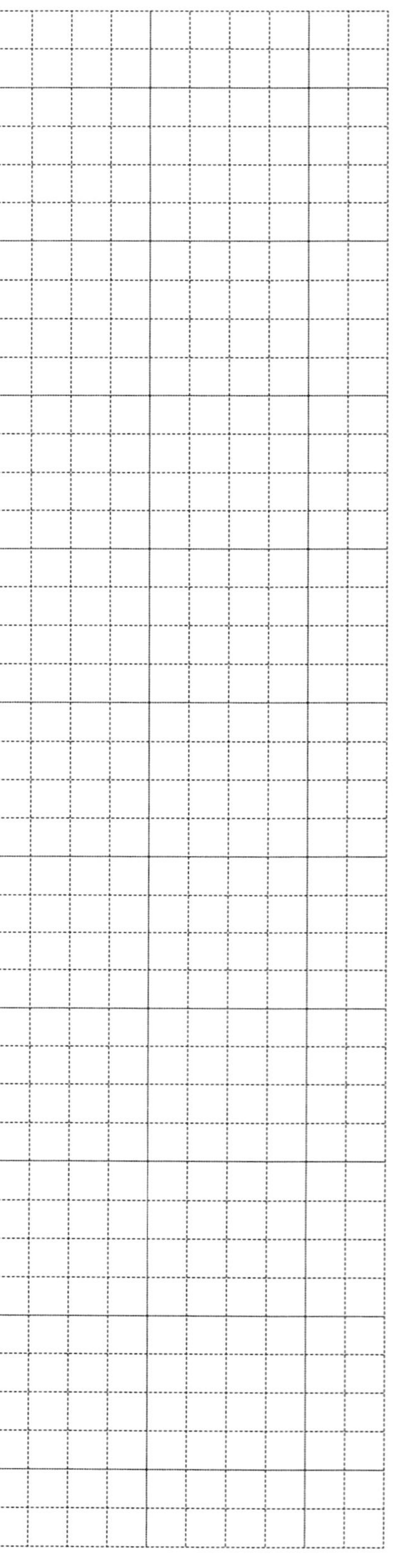

negative value, because even without stimulation there is some A current (*gA* becomes positive, lower window). Now turn on the stimulator (1 nA) and observe that the impulses begin after a delay of about 0.5 s. This delay is a result of the large, outward (positive) A current that is activated by the depolarization and that keeps the membrane potential from exceeding threshold. Because the A conductance inactivates, its amplitude decreases with time (watch *gA*) and consequently no longer acts to hold the membrane potential below impulse threshold. (The regular spikes in the A conductance trace are caused by the soma membrane potentials associated with the axon impulses.)

Set the stimulator to "off" and increase the stimulus amplitude to 2 nA. Now when you turn on the current (stimulator to "on"), impulses are generated almost immediately; however, the initial frequency is low. As time goes on, *gK* again inactivates, causing the impulse frequency to increase until a steady firing frequency is achieved. Now reduce the current to 1 nA. Observe that the impulse frequency slowly decreases to a new steady-state level. Evidently the overall effect of the A current is to reduce the rate at which impulse frequency changes in response to step changes in inputs.

Neuron-4 Postinhibitory rebound (PIR) due to g_h

For this exercise NEURON again is configured to simulate a two-compartment neuron with a soma and an axon. Here, however, the soma includes the h current, a time-dependent inward current (at normal neuronal membrane potentials) that is activated by hyperpolarization. Note that this differs from the A current, which both activated and inactivated on depolarization. Like the A current, the activation characteristics of the h current are much slower than those of the impulse-related currents. When you begin this exercise the computer screen displays a TIMESERIES window that displays the same four windows open in Neuron-3, except that *gh* rather than *gA* is graphed in the lower window together with the axon membrane potential (*VmA*, fig. I.5-5).

The aims of this exercise are 1) to illustrate how *gh* generates a sag potential that counteracts membrane hyperpolarization when current steps are applied to the soma and 2) to demonstrate how *gh* leads to postinhibitory rebound following hyperpolarizing current pulses and how it leads to negative after-potentials following depolarizing pulses.

Begin the simulation by clicking on the STOP/GO icon, ignoring the first 3 s of the graph. (Transients which do not reflect the true membrane potentials of soma and axon occur at this time.) Observe that the resting potential in this simulation is about -60 mV, rather than -70 mV as in the previous simulations. This depolarization is due to the inward current through h channels, which are partially activated at rest. (You can verify this statement by setting *gh* to zero.) Now turn on the stimulator to inject a hyperpolarizing current of 2 nA. Note that the membrane potential becomes more negative for a short interval but then sags back a few mV towards the resting level. This upward swing in the potential is caused by the activation of the inward current I_h (caused by the hyperpolarization-induced increase in *gh*, lower window).

Turn the stimulator off after about 0.5 s of simulated time, when the trace

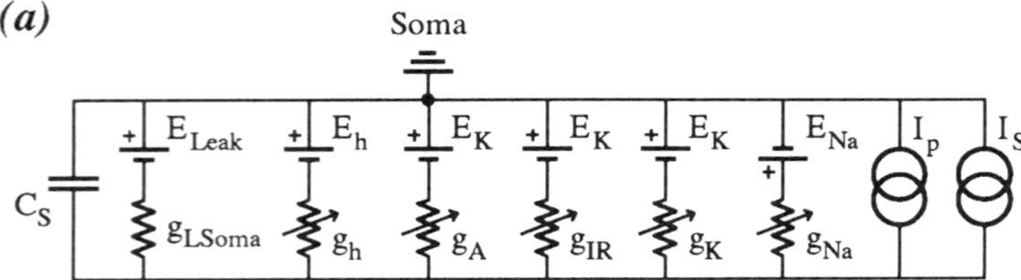

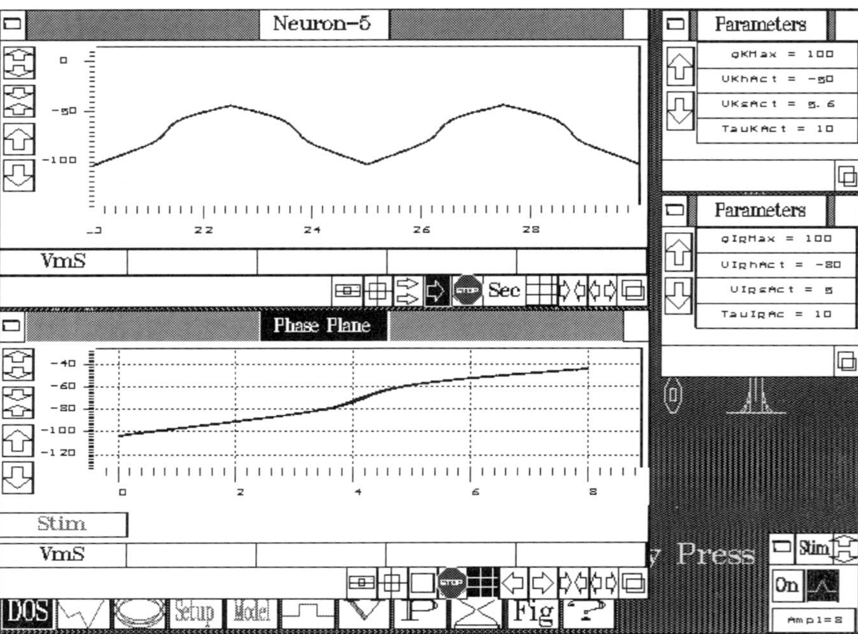

Figure I.5-6. One-compartment NEURON model. *a*) Reduced one-compartment equivalent circuit for NEURON. *b*) Monitor display for exercise Neuron-5

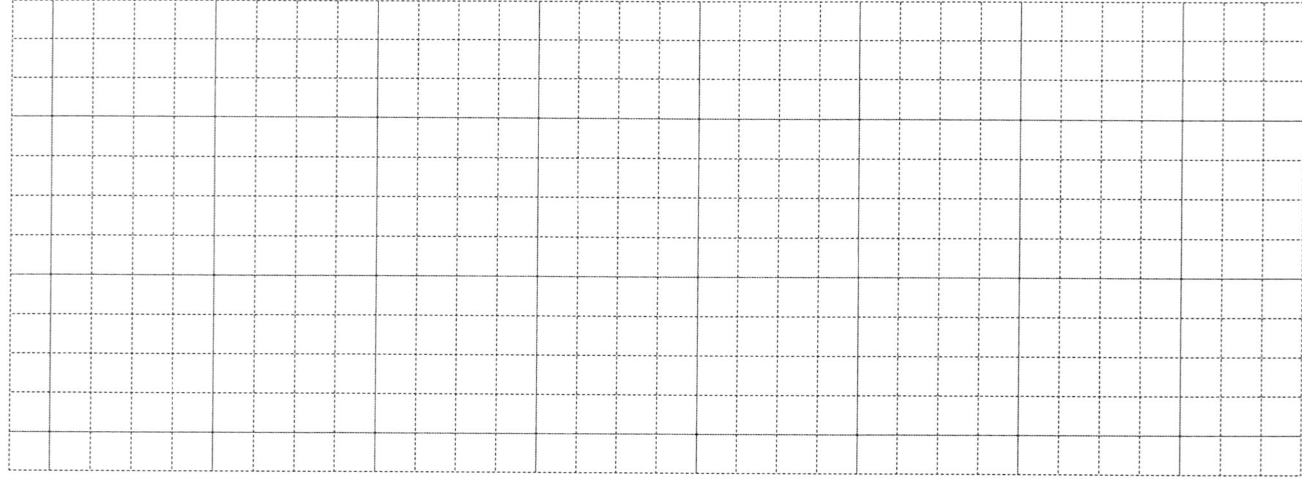

is near the center of the window. The soma potential (upper window) rapidly returns to the resting level and then continues to depolarize beyond rest. This depolarization is coupled to the axon compartment, which responds with a series of rapidly repeating impulses. These slow down and then stop as the membrane potentials in soma and axon return to their resting levels. The cause of this postinhibitory rebound is immediately apparent from an inspection of the h conductance trace (*gh*) in the lower window. Hyperpolarization leads to an increase in the h conductance, this conductance deactivates slowly when the current pulse ends. Decay of PIR coincides with, and is caused by, the deactivation of the h conductance.

Set the stimulator to inject a long (about 10 s) 2 nA depolarizing current into the soma. This induces a large depolarization and leads to rapid impulse activity. During this interval, the h channel deactivates very slowly. Now turn the current off and notice that the soma membrane potential hyperpolarizes below the resting level. Explain why. Such after-potentials are observed in many neurons. Explore both PIR and the hyperpolarizing after-potential by applying current pulses with a range of amplitudes and durations. Very brief or low-amplitude current pulses generate almost no after-potentials. Why?

Neuron-5 Nonlinear *I-V* curves

For this exercise NEURON is configured to simulate a reduced, one-compartment neuron with a soma, but no dendrite or axon (fig. I.5-6a). This simulation is equivalent to an extended neuron in which all of the components are at the same potential. The only compartment simulated, the soma, is similar to the soma compartment of the three-compartment model simulated in exercise Neuron-1. In this exercise, however, two voltage-dependent conductances are included. The first is the delayed rectifier conductance (g_K), which activates on depolarization to permit the efflux of potassium ions. It is very similar to the delayed rectifier conductance that functions to rapidly terminate the nerve impulse. The second conductance (g_{IR}) known as the "inward rectifier" activates when the membrane potential is hyperpolarized below the potassium equilibrium potential (below −80 mV in this simulation). At the beginning of the exercise the computer screen displays five windows: a TIMESERIES that displays the soma membrane potential (*VmS*); a PHASEPLANE window that displays the soma membrane potential as a function of injected current (*Stim*); two PARAMETER windows; and the STIMULATOR window (fig. I.5-6b). The aim of this exercise is to illustrate the shape of an automatically generated *I-V* curve in a cell with two potassium conductances, the delayed rectifier (activated by depolarization) and the inward rectifier (activated by hyperpolarization).

When you begin the exercise, the stimulus current is directed to the soma with a full-scale value of 8 nA. Because there is also a tonic (d.c.) current of −4 nA applied via parameter *Idc*, the actual change of the potential on the abscissa of the PHASEPLANE window is −4 nA to +4 nA. Click on the STOP/GO icon to initiate the graphing. As the trace moves to the right, the membrane is slowly depolarized by the linear ramp of injected depolarizing current generated by the stimulator. Because of the linear relationship between current amplitude and time (the duration of the full triangle current waveform is 5.0 s), the graph plotted on the upper window of *Vm* against time has the same shape as the waveform in the lower window, where *Vm* is graphed against current amplitude. In both graphs the membrane potential is at −70 mV (rest) when no current is injected into the soma. Note the sigmoidal shape of the *I-V* graph. This shape is typical of many real neurons because of the presence of multiple voltage-sensitive channels in dendrites and neuronal somata.

To explore how the delayed rectifier and inward rectifier potassium conductances affect the *I-V* curve, compare the graphs generated by setting first the delayed rectifier conductance (*gKMax*) and then the inward rectifier conductance (*gIR*) to zero. Open up the appropriate PARAMETER MODIFICATION windows to make these changes in parameter values. Finally set both of these voltage-dependent conductances to zero and compare the straight line generated in the lower window to previous graphs.

Section I.6 Electrophysiology of Neuronal Interactions

I.6.1 INTRODUCTION

Two aspects of electrophysiology are critical for the functioning of the nervous system—the properties of individual neurons, which we have just described, and the properties of synaptic interactions, described in this and the following sections. The cell doctrine that animal tissues are constructed from individual cells applies fully to neurons as well. Each neuron is an individual unit, separated from other neurons but linked to them by special points of contact, the synapses (a term coined by Sherrington near the turn of the twentieth century). In this section we present an introduction to the electrophysiology of neuronal interactions, including both chemical synapses and electrical junctions. The former are sites where unidirectional flow of information occurs, from the pre- to the postsynaptic neuron. Such chemical interactions often include some amplification and are subject to extensive short- and long-term modulation by neurohormones. Electrical junctions, on the other hand, invariably lack amplification but act as very fast, rather rigid, often bidirectional connections.

The initial experiments demonstrating the existence of chemical synaptic transmission were carried out by Otto Loewi in 1921. He showed that a "vagus substance" (later found by him to be acetylcholine—ACh) mediates the slowing of the frog heart rate following stimulation of the vagus nerve. In 1936, Henry Dale demonstrated that ACh is the chemical transmitter by which motor neurons excite muscle fibers at the neuromuscular junction. We now know that most neuronal interactions in higher animals occur because of the release of a wide variety of neurotransmitter substances. Interaction of the transmitter molecules with specific receptor proteins in postsynaptic membrane induces membrane potential changes in the postsynaptic cell.

I.6.2 SYNAPTIC POTENTIALS

Very briefly, the sequence of events occurring during synaptic transmission begins when depolarization of the presynaptic nerve terminal opens voltage-gated calcium channels. The increased intracellular calcium concentration in the nerve terminal of the presynaptic neuron mediates, via a series of steps, the release of neurotransmitter from synaptic vesicles. The transmitter molecules diffuse across the synaptic cleft, the 20–40 nm gap that separates pre- and postsynaptic membranes, to bind selectively to receptor molecules embedded in the membrane of the postsynaptic cell. The transmitter-receptor interaction is often very brief because 1) degradative enzymes found in the cleft deactivate the transmitter, 2) transmitter dilution occurs by diffusion, and 3) sequestration mechanisms such as transmitter uptake back into the presynaptic terminals actively reduce the concentration of transmitter in the cleft. For the direct-acting synapses described in this section, the receptor protein in the postsynaptic membrane are channels much like the voltage-gated channels described earlier. These are ion channels opened by transmitter binding. Ions entering or exiting the postsynaptic membrane create transmembrane currents that cause changes in membrane potential.

The summation of ionic currents through hundreds or thousands of activated receptor protein molecules induces the change in membrane potential known variously as the synaptic potential, the junctional potential (at the neuromuscular synapse), or the postsynaptic potential. More specifically, the postsynaptic membrane potential response is known as an "epsp" (excitatory postsynaptic potential) if the synaptic potential is excitatory or as an "ipsp" (inhibitory postsynaptic potential) if the synaptic interaction is inhibitory. The final determinant of whether a given synaptic interaction is excitatory or inhibitory is whether the interaction tends to increase (excitatory) or decrease (inhibitory) the impulse frequency of the postsynaptic neuron. For vertebrate striated muscles, in which each muscle cell receives synaptic input from a single motor neuron, the synapses act as an amplifying relay; each presynaptic impulse gives rise to a very large epsp (a.k.a. an "ejp," excitatory junctional potential) that elicits a single postsynaptic impulse. Neurons, on the other hand, often receive synaptic inputs concurrently from hundreds or thousands of presynaptic neurons. For neurons, then, the impulse rate depends on the sum of all of these synaptic inputs. This sum is computed at the axon hillock, the integrative sector of neurons from which impulses arise. The weights of individual synaptic inputs depend on the specific morphology of the postsynaptic neuron (see modeling exercise Synapse-1).

I.6.3 SHAPE OF SYNAPTIC POTENTIALS

There is much variety in the shapes and amplitudes of synaptic potentials. They may be depolarizing or hyperpolarizing, of brief or of long duration. At some neuronal synapses synaptic potentials are caused by the release of only a few quanta (vesicles-full) of transmitter molecules; and therefore they may be tiny, less than a 1.0 mV in amplitude. At the neuromuscular junction, by way o contrast, concurrent release of hundreds of quanta may cause ejp's that are more than 50 mV in amplitude (nearly half the size of a nerve impulse). Brief synaptic potentials observed in many synapses are characterized by a rapid change in potential that rises to a peak value within several milliseconds. Decay from this peak then occurs exponentially with a time constant of about 10 ms. For chemical synapses there is a delay between the presynaptic depolarization and the onset of the postsynaptic response. For the fastest acting synapses this delay is about 0.5 ms.

I.6.4 SYNAPTIC SUMMATION

Neurons usually receive barrages of synaptic inputs that, combined with their resting properties, determine the membrane potential and the impulse frequency. Two terms are used to describe the effects of this multiple input. The term "temporal summation" refers to the summing of synaptic potentials at a single synapse. If a presynaptic neuron has a relatively high impulse frequency, the synaptic potentials can overlap and sum. At inhibitory synapses temporal summation leads to greater and longer inhibition than would be achieved by a single synaptic potential; whereas at excitatory synapses larger, prolonged excitatory responses would occur. The term "spatial summation" refers to the combined potentials resulting when several presynaptic cells are simultaneously active. Again, when synaptic inputs occur at nearly the same time, the effect of spatial summation is to produce greater and more prolonged excitation or inhibition. When combinations of excitatory and inhibitory inputs occur, the overall effect is for the inhibition to cancel some of the excitatory effects. Both temporal and spatial summation at fast synapses are the result of charge storage by the membrane capacitor during synaptic transmission. A heuristic approach is to think of synaptic inputs as brief current pulses that are injected into postsynaptic neurons through the action of the presynaptic cell. These currents are applied to the membrane capacitor, where they add in a nearly linear fashion. Synaptic currents that occur before the capacitor has discharged from the previous input add algebraically to the charge already present. It should be noted for completeness that synaptic potentials, which are defined by the effect of a presynaptic neuron on its target, can occur also if the action of the neurotransmitter is to close an ionic channel. Such conductance decrease synapses often act slowly over several seconds (see modeling exercise Synapse-2).

I.6.5 THE PARALLEL CONDUCTANCE MODEL OF SYNAPTIC TRANSMISSION

We can understand the electrophysiology of synaptic transmission by returning to the parallel conductance model, which has served so well as an aid to understanding nerve impulses. To describe synaptic interactions we need to introduce some new conductances and batteries into the basic model (fig. I.6-1). Let g_s be the conductance associated with the opening of a set of ligand-gated (i.e., neurotransmitter-activated) ion channels, and let E_s be the potential of a battery associated with that specific set of channels. The term E_s is called the "reversal potential" for receptor conductance for reasons that will be apparent soon. The current through an ensemble of channels is given by

$$I_s = g_s * (V_m - E_s). \qquad \text{(I.6-1)}$$

In this equation, g_s is nonzero only briefly, while the transmitter molecules are bound to and open the channel. Thus the synaptic current, which is zero unless g_s is nonzero, also is nonzero only briefly during synaptic transmission. The specific value of this synaptic conductance varies with the size of the synaptic contact; the number of vesicles released, which may vary with presynaptic properties such as the amplitude and duration of the presynaptic depolarization; previous activity in this synapse, for example, the state of synaptic fatigue; and postsynaptic properties, such as the density and number of receptor molecules. Remember that, like voltage-gated channels, the individual ligand-gated channels have two primary states, open and closed. The value of g_s may be computed from the number of channels open at any time multiplied by the single-channel conductance.

I.6.5.1 Reversal potential

The term $(V_m - E_s)$ in equation (I.6-1), the difference between the membrane potential of the postsynaptic neuron and the receptor reversal potential, is the electrochemical driving force for ion flow through the receptor channels. Depending on the relative values of the two potentials in this term, the driving force can be positive, negative, or zero; however, it is independent of the channel conductance. The synaptic battery potential, E_s, is usually referred to as the reversal potential for a given synapse because as V_m is varied the synaptic current reverses sign when V_m is equal to E_s. So, when V_m is more negative than E_s, the current is negative (i.e., inward and therefore depolarizing); whereas, when V_m is less negative than E_s, the current is positive (i.e., outward and hyperpolarizing). Outward, hyperpolarizing synaptic currents are always inhibitory; whereas inward, depolarizing currents are often, but not always, excitatory.

The specific value for the reversal potential for a given synaptic current is determined by the ion selectivity of the receptor channels. Inhibitory interactions often involve the activation of chloride or potassium channels. Because these channels are selectively permeable to a single ion species, their reversal potentials are the equilibrium potentials (Nernst potentials) for chloride and potassium ions, respectively. Other channels, such as the nicotinic ACh

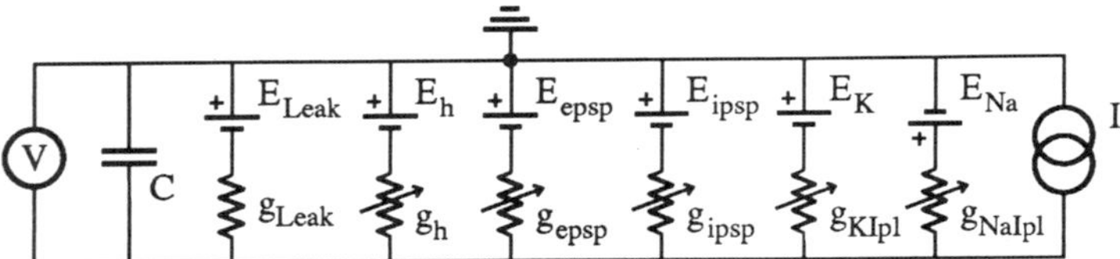

Figure I.6-1. Equivalent circuit for the CIRCUIT model

receptor, are less selective, with significant conductance for several ions. The nicotinic ACh channel, for example, is nearly equally permeable to sodium and potassium ion. The reversal potential for this channel therefore lies nearly halfway between the Nernst potentials for these two ions, about -10 mV in many cells. Determining the reversal potential for a given synaptic conductance provides one important clue to the identity of the receptor responsible for the conductance and also to the identity of the neurotransmitter that activates the conductance (see modeling exercise Synapse-3).

I.6.5.2 Relationship between synaptic current and synaptic potentials

We can now understand the shape of many synaptic potentials. Note that the parallel conductance model of figure I.6-1 includes both a capacitor and a "resting" nonsynaptic conductance. Activation of a synaptic conductance in this model generates a synaptic current. At synapses with rapid, brief responses, some fraction of this pulse-like synaptic current alters the charge on the capacitor, which does not directly change the membrane potential. The remaining current passes through the resting conductance and acts to depolarize or hyperpolarize the membrane. Thus the amplitude of the synaptic potential will depend on the amplitude of the synaptic current, the size of the membrane capacitance (nearly 1.0 μF/cm^2 for all cell membranes), and the value of the resting conductance. Qualitatively, the amplitude will be relatively large if the resting conductance is small and small if the resting conductance is large. This is not, in fact, the whole story. The rapid, rising phase of a synaptic potential represents the early, exponential charge or discharge of the membrane capacitor. Because synaptic currents are often very brief, the charge on the capacitor never reaches a steady level. The approach to steady state increases as the duration of the synaptic current is increased; hence, synaptic potentials become relatively larger when the duration of the synaptic current is extended.

Following the cessation of the synaptic current, the membrane potential declines exponentially with a time-constant that is the product of the membrane capacitance and the resting conductance. Thus for synaptic potentials caused by brief changes in synaptic conductance the decay of the synaptic potential reflects the properties of the postsynaptic cell, not those of the presynaptic neuron. In some synaptic interactions, however, the membrane potential rises and decays gradually (over the course of hundreds of ms). For these slow interactions the capacitative current is nearly zero, and we may safely ignore the capacitor. The equation for the postsynaptic membrane potential at slow synapses is given by

$$V_m = (g_r E_r + g_s E_s)/(g_s + g_r), \tag{I.6-2}$$

where all of the terms are those already encountered earlier. An easy way to remember this formula is to note that the membrane potential during slow-acting synaptic transmission is the weighted sum of the resting and synaptic battery potentials. The weighing factor is the ratio of the resting and synaptic membrane conductances to the total conductance ($g_T = g_s + g_r$). In these terms,

$$V_m = (g_r/g_T) * E_r + (g_s/g_T) * E_s. \quad (I.6\text{-}3)$$

Note that the synaptic conductance increases the total conductance of the neuron (see modeling exercise Synapse-4).

I.6.6 SYNAPTIC FATIGUE

Our conceptual picture for synaptic transmission is that of a presynaptic terminal replete with synaptic vesicles whose contents are rapidly dumped into the synaptic cleft when an impulse arrives at the presynaptic terminal. Actually, only a small number of vesicles may be poised "mobilized" for immediate release. When this mobilized transmitter is depleted, some time is required to mobilize additional vesicles. If a second impulse arrives at the presynaptic terminal while transmitter mobilization is incomplete, few vesicles will be available, and consequently the second postsynaptic response will be smaller than the first. The phenomenon described by this scenario is called "synaptic fatigue." This conceptual model explains why at some synapses the second of two closely paired synaptic potentials is smaller than the first. During a train of presynaptic impulses, fatiguing synapses respond with a series of diminishing synaptic potentials. Synaptic fatigue is a short-term phenomenon; after a few seconds of presynaptic inactivity, the synaptic potentials recover to their full amplitude (see modeling exercise Synapse-5).

I.6.7 CHEMOTONIC INTERACTIONS

The release of transmitter from presynaptic vesicles is intimately tied to an increase in the calcium concentration in the presynaptic terminal. Any process that raises the concentration of calcium in the terminal, even injection of calcium through a micropipette, will induce the release of transmitter. As describe above, voltage-gated calcium channels, which are opened by the large depolarization characteristic of nerve impulses, provide pores for calcium influx in the presynaptic terminal. Because calcium-channel gating depends on membrane depolarization, not on an impulse per se, any signal that depolarizes the terminal is able to induce transmitter release and cause synaptic transmission. Direct, non–spike-mediated control of transmitter release occurs in many neurons, including those that are incapable of generating impulses. Such neurons include the photoreceptors and bipolar cells in vertebrate eyes and a variety of neurons in the invertebrates. In these neurons, calcium channels in the synaptic terminals may be partially activated at the resting potential, causing a continuous release of transmitter. A depolarization in such a cell will open more channels (strictly speaking, increase the probability that channels will be open) and therefore increase transmitter release. Conversely, a hyperpolarization imposed on this cell will decrease calcium influx and hence decrease the transmitter release. This example illustrates "chemotonic" synaptic transmission, in which the control of transmitter release by the presynaptic potential alters the postsynaptic current without the need for nerve impulses. Chemotonic synaptic transmission is effective only if the presynaptic terminal is not very far removed, electrically speaking, from the integrative regions, such as in vertebrate retinae It might be useful to think of direct, non-spike-meditated synaptic transmission as "analog" synaptic transmission, rather than "digital" signaling (see modeling exercise Synapse-6).

I.6.8 ELECTROTONIC INTERACTIONS

A great controversy raged during the first half of the twentieth century about whether neurons interact electrically or via chemical transmitter substances. Although the polemic was fruitful, it was also based on a false dichotomy. Neurons interact not only by chemical synapses, as we have just seen, but also electrically via gap junctions. Gap junctions are formed from the close apposition of protein pores in two neighboring neurons. Because these pores are large, they allow ions and even small molecules (such as marker dyes) to pass nonselectively between cells. Many gap junctions are like electrical resistors between two neurons; currents pass from cells with less negative membrane potential to those where the potential is relatively greater (more negative). The effect of these currents is to reduce the potential difference between electrically coupled neurons. Although there is a huge range of interaction strength at electrical junctions, a typical coupling coefficient is 0.1, meaning that a change of 10 mV in one neuron will cause a change of 1 mV in the neuron to which it is connected by an electrotonic interaction.

In addition to their acting simply as pores between two neurons, electrical junctions may be more complex. One example is that of rectifying electrical junctions, where the channel conductance is controlled by the potential difference between the two neurons. Current will pass in one direction through such junctions but not in the reverse. Such behavior is similar to that seen in electrical rectifiers or diodes. Rectifying interactions characterize some electrical synapses, where the junctions between neurons are organized into pre- and postsynaptic elements that are functionally, though not morphologically, similar to those found at chemical synapses. At electrical synapses, the large depolarization caused by a presynaptic impulse induces a large, brief current flow through the junction to briefly depolarize the postsynaptic neuron. Unlike the situation at chemically mediated synapses, there is little or no synaptic delay at electrical synapses (see modeling exercise Synapse-7).

(a)

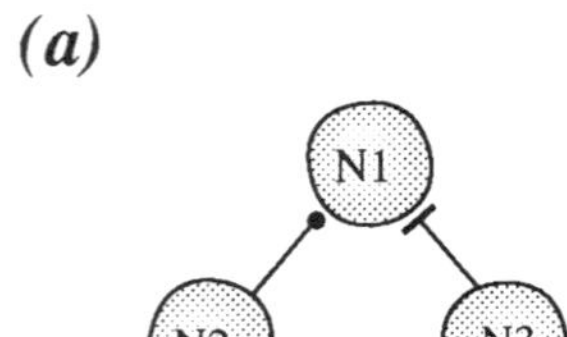

Figure I.6-2. Inhibitory and excitatory interactions. *a*) Diagram of synaptic interactions: N2 inhibits N1; N3 excites N1. *b*) Monitor display for exercise Synapse-1

(b)

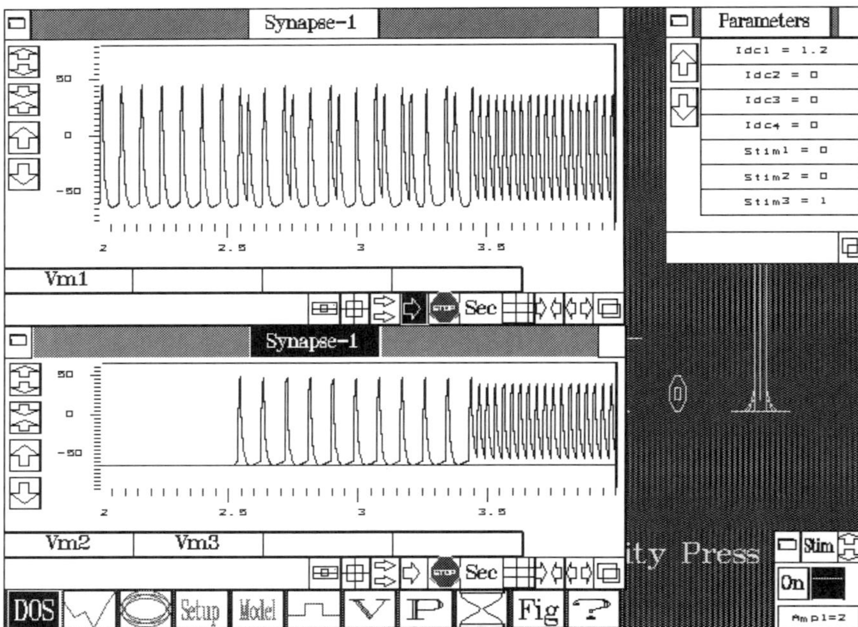

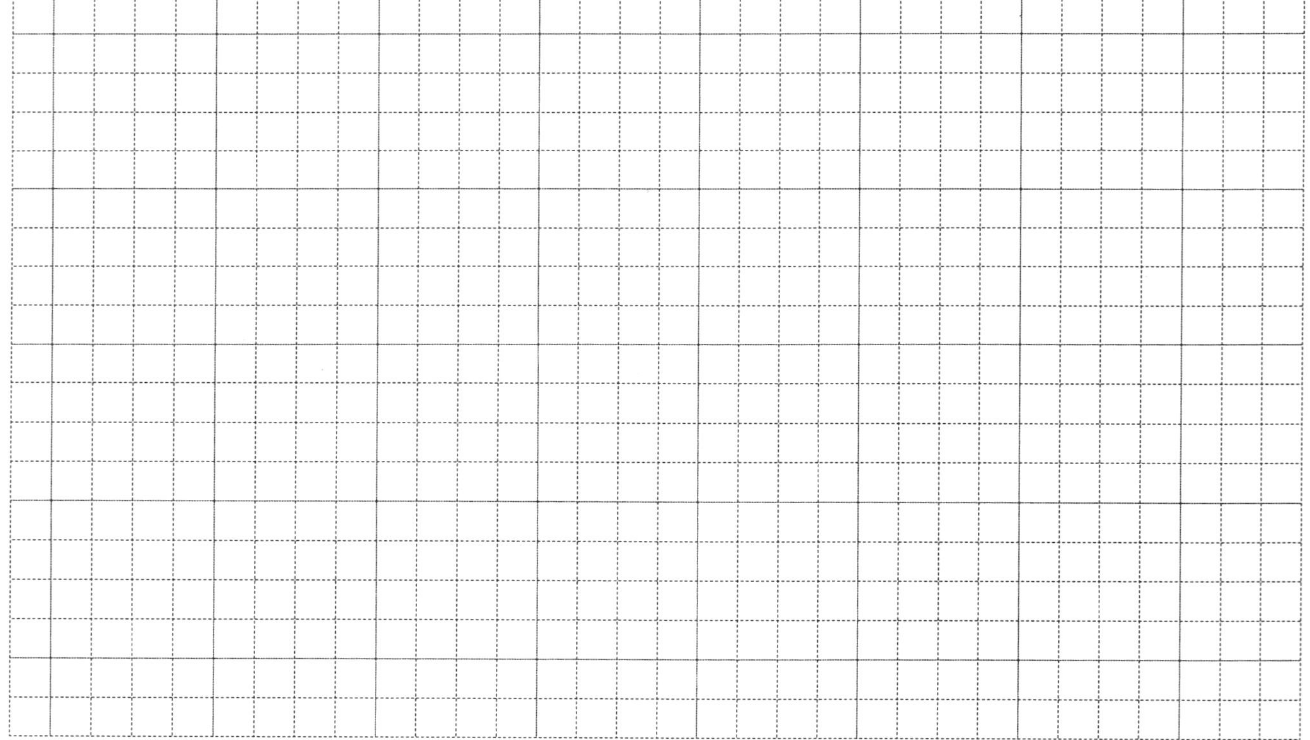

Modeling Exercises

NeuroDynamix **Model: CIRCUIT** **Exercises: Synapse-1 to to Synapse-7**

Introduction

This series of modeling exercises is designed to help you to understand the physiology of synaptic transmission between neurons. Three types of connections between neurons are demonstrated: chemical excitation, chemical inhibition, and electrical interaction. To carry out these exercises you will be using the CIRCUIT model. Please refer to section II, "Description of the Models," for a description of the specific variables and parameters associated with CIRCUIT. Remember that the simulated neurons are isopotential; i.e., they are treated as one-compartment neurons. After you have completed each of the exercises outlined here, feel free to alter model parameters, such as the synaptic amplitude and time course, to get a better feel for how neurons interact.

Synapse-1 Functional role of synaptic interactions

For this exercise CIRCUIT is configured so that two presynaptic neurons, one excitatory, and the other inhibitory, control the membrane potential and therefore the impulse frequency (firing rate) in a postsynaptic neuron (fig. I.6-2a). Four windows (fig. I.6-2b) are open when you begin this exercise: two TIMESERIES windows, one that graphs the membrane potential (*Vm1*) of the postsynaptic neuron (N1) and another that graphs the potentials (*Vm2* and *Vm3*) of the inhibitory (N2) and excitatory (N3) presynaptic neurons, respectively; the PARAMETER window, for controlling the potentials of all three neurons; and the STIMULATOR window. The purpose of this exercise is to demonstrate the functional role of excitatory and inhibitory inputs in determining the impulse rate of a postsynaptic neuron.

When this exercise begins (click on the STOP/GO icon), the postsynaptic cell (N1, upper window) is depolarized slightly (*Idc1* = 1.2 nA) to generate a low-frequency train of impulses; the presynaptic neurons are silent. Now click the right mouse button on *Stim3* to connect the stimulator to the excitatory neuron (N3). Activate this cell by setting the stimulator amplitude (*Ampl*) to 1 nA. The excitatory input from N3 nearly doubles the firing rate of N1. By raising the firing rate of N3 to higher levels, the firing rate of the postsynaptic neuron increases still further. Return the system to its initial state by setting the stimulator amplitude and *Stim1* to 0. Continue this exercise by inhibiting the activity in N1 by stimulating the presynaptic cell N2. (Set *Stim2* to 1 and the stimulator current to 1 nA.) Now the inhibitory potentials evoked in N1 by N2 reduce the firing rate of this postsynaptic cell. If you increase the firing rate of the inhibitory neuron still further, impulses are completely suppressed in N1.

Make a table listing the impulse frequency of N1 when the presynaptic

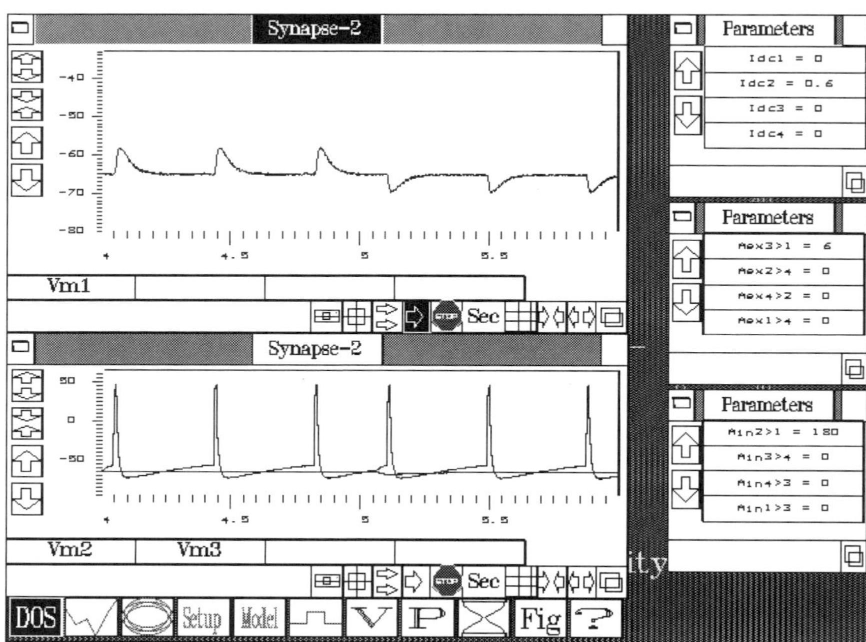

Figure I.6-3. Monitor display for exercise Synapse-2

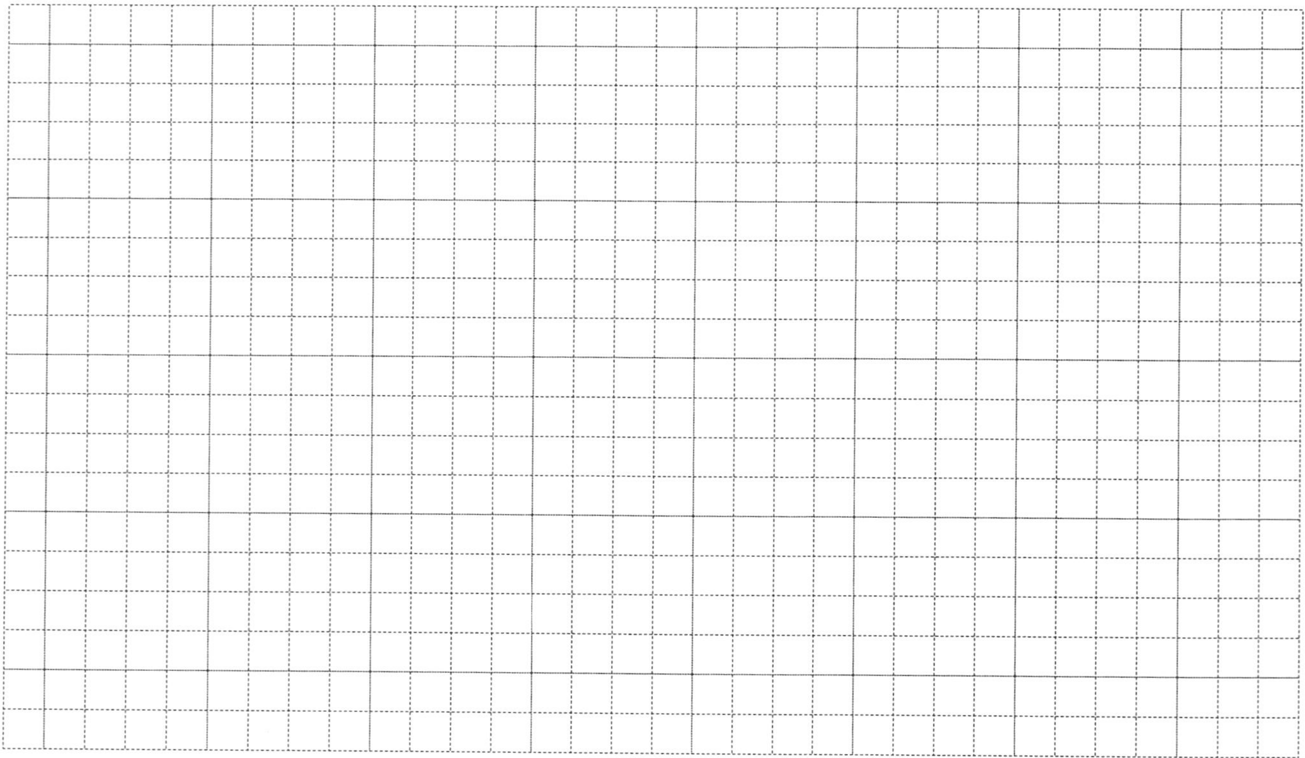

neurons are silent, when these cells are activated individually, and when the presynaptic neurons are driven simultaneously with 1 and 2 nA currents. (This table can be generated readily by opening PARAMETER MODIFICATION windows for parameters *Stim2* and *Stim3* and turning the stimulator off.) Your table should help you to understand that the impulse activity in neurons is controlled by the weighting of the activity in presynaptic excitatory and inhibitory neurons. Explore the relationships between presynaptic activity and postsynaptic responses more carefully by using the PARAMETER MODIFICATION windows for precise control of presynaptic cell firing rates.

Synapse-2 Excitatory and inhibitory synaptic potentials

For this exercise CIRCUIT is configured, as in Synapse-1, so that two presynaptic neurons, one excitatory and the other inhibitory, generate synaptic potentials in a postsynaptic neuron (fig. I.6-2a). Five windows (fig. I.6-3) are open when you begin this exercise. Two are TIMESERIES windows, one that graphs the membrane potential (*Vm1*) of the postsynaptic neuron (N1) and another that graphs the potentials (*Vm2* and *Vm3*) of the inhibitory (N2) and excitatory (N3) presynaptic neurons, respectively. The other three are PARAMETER windows, one for controlling the potentials of all three neurons, the other two for setting the synaptic strengths of the excitatory and inhibitory synapses, respectively. Some of the model parameters have been altered slightly from those of exercise Synapse-1. In particular, the tonic excitation has been removed from the postsynaptic neuron N1 so that this cell is not spiking. The purpose of this exercise is to illustrate the shapes of postsynaptic excitatory and inhibitory potentials, epsp's and ipsp's, respectively.

After you have initiated the exercise by clicking on the STOP/GO icon, activate the presynaptic excitatory neuron (N3) to generate impulses at a low rate by setting *Idc3* to about 0.6 nA. Notice that each presynaptic impulse evokes an epsp that is characterized by a rapid rise and a relatively slow decay. This is the characteristic shape of epsp's at direct-acting synapses. Remove the excitatory current from N3 and set *Idc2* to 0.6 nA. Now the slow firing of N2 induces synaptic potentials, ipsp's, in N1 that are the mirror image of the epsp's evoked by N3. Change the amplitude of these synaptic potentials by varying the parameters *Aex3* > *1* in the middle PARAMETER MODIFICATION window and *Ain2* > *1* in the bottom window. When the amplitude of the epsp is increased you will note that impulses are triggered by each presynaptic impulse (tops of spikes are clipped by the graph). Notice that although the epsp amplitude can vary considerably, the ipsp amplitude cannot be made very large. Why? [Hint: Compare the values of the resting potentials for the neurons (*ELeak*) and the reversal potentials for the synaptic potentials (*Eepsp* and *Eipsp*) and see Synapse-3.]

(a)

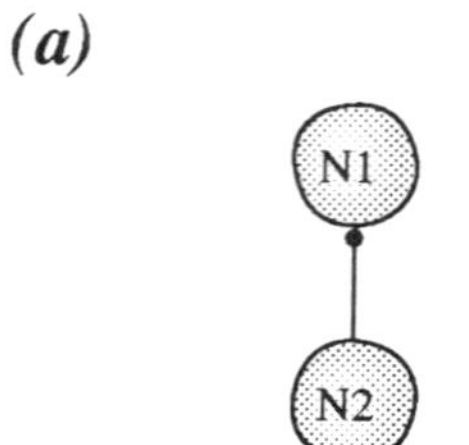

Figure I.6-4. Synaptic inhibition. *a*) Diagram of synaptic interactions: N2 inhibits N1. *b*) Monitor display for exercise Synapse-3

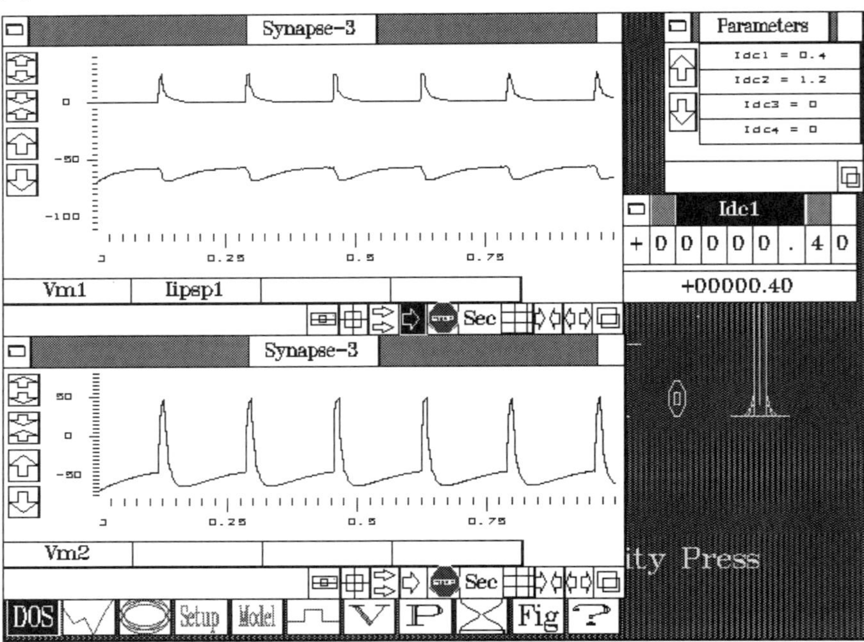

Synapse-3 Reversal potential for inhibitory synapses

For this exercise CIRCUIT is configured to simulate two neurons: an inhibitory presynaptic neuron, N2 (*Vm2*, lower window, fig. I.6-4a, b), connected to a postsynaptic cell, N1 (*Vm1*, upper window). The model is configured initially so that N2 fires at a constant rate to generate a series of ipsp's in N1. The PARAMETER window is open so that you can control the neuronal membrane potential of the postsynaptic neuron via current injection. The purposes of this exercise are to demonstrate the relationship between synaptic currents and potentials and to illustrate the phenomenon of reversal potential. These procedures are of some physiological importance because synaptic interactions are identified in part by the postsynaptic potential at which the synaptic current reverses.

Begin this exercise by first opening the PARAMETER MODIFICATION window for *Idc1*; this will allow you to manipulate the membrane potential of N1 easily. After you initiate graphing by clicking on the STOP/GO icon, observe first the relationship between the presynaptic impulses (*Vm2*) and the postsynaptic responses. The current pulses generated by the opening of synaptic channels (simulated chloride channels) in N1 (*Iipsp1*, upper window) generate the negative-going synaptic potentials in this cell (*Vm1*, upper window). Compare the duration and time course of the synaptic currents and membrane potential excursions. This exercise simulates a fast, direct-acting synapse where the synaptic current (outward) has a shorter duration than the postsynaptic response. Now increase *Idc1* in 0.1 nA steps to gradually depolarize the postsynaptic cell. Why do the amplitudes of both synaptic current and the ipsp increase in size as N1 is depolarized? Now reset *Idc1* to zero and step the current to small negative values. Observe that the synaptic current and ipsp are diminished in size as the injected current is increased. Continue to increase the amplitude of the (negative) current until both the postsynaptic current and the ipsp disappear. Note that the value of the membrane potential of N1 then increases the hyperpolarizing current injected into N1 still further. Now the directions of both the current and the synaptic potential are reversed! The synaptic potential is now in the depolarizing direction! Why? The postsynaptic potential at which the synaptic response disappears is called the reversal potential because the synaptic response changes sign at this potential. Determining the reversal potential is one very important step for identifying the ions carried by synaptic channels. A further step in the identification of synaptic channels is to observe changes in the reversal potential caused by altering ionic concentrations in the physiological saline, that is, by changing the equilibrium potentials of specific ions.

Synapse-4 Relationship between synaptic currents and potentials

For this exercise CIRCUIT is configured, as in Synapse-1, with two presynaptic neurons, of which N2 is excitatory and N3 is inhibitory. Both of these neurons induce synaptic currents that lead to synaptic potentials in the postsynaptic neuron N1 (fig. I.6-2a). Five windows (fig. I.6-5) are open when you begin this

Figure I.6-5. Monitor display for exercise Synapse-4

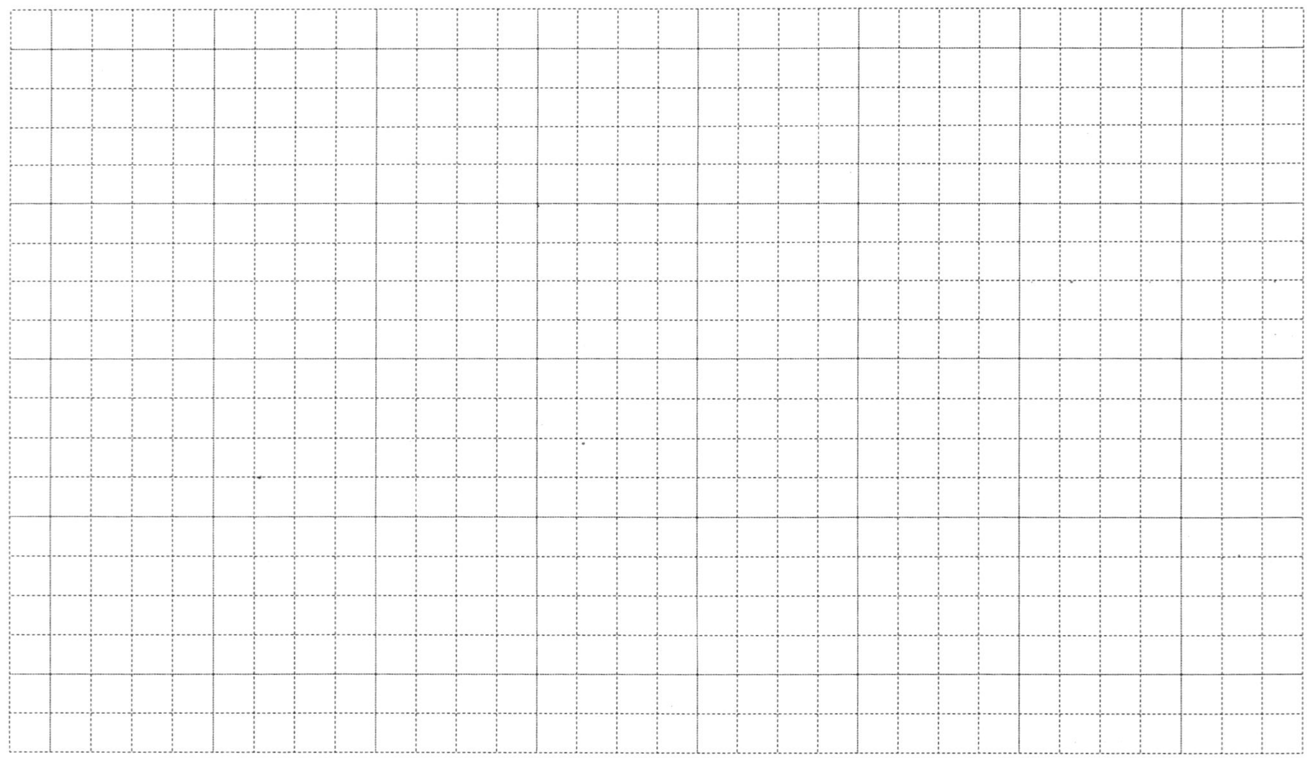

exercise. Two are TIMESERIES windows. The upper one graphs the membrane potential (*Vm1*) of the postsynaptic neuron and the excitatory and inhibitory synaptic currents (*Iepsp1* and *Iipsp1*), whereas the lower window graphs the potentials (*Vm2* and *Vm3*) of the excitatory (N3) and inhibitory (N2) presynaptic neurons, respectively. There are three PARAMETER windows open, the top one for controlling the membrane potentials of all three neurons, the middle one for setting the synaptic strength for the excitatory synapse, and the bottom one for setting the strength of the inhibitory synapses. The purposes of this exercise are to demonstrate the relationships between synaptic currents and synaptic potentials and to demonstrate the interactions of synaptic potentials in spatial and temporal summation.

Initiate this exercise by clicking on the STOP/GO icon, then set *Idc3* to generate impulses at a low rate in N3. Note that each impulse in N3 is followed by a negative deflection in the *Iepsp1*. This brief, negative, inward current results from an increased conductance in the excitatory synapse between N3 and N1. These current pulses then depolarize the membrane to generate longer-lasting epsp's (*Vm1* trace, upper window). Verify that this interaction is indeed excitatory by increasing the depolarizing current applied to N2. Note that as the impulse frequency increases, the epsp's in N1 begin to sum, depolarizing N1 continuously. This "temporal summation" occurs, even though the current pulses do not overlap, because of the summation performed by the cell membrane capacitance.

In order to study the relationship between the membrane current and the synaptic potentials at an inhibitory synapse, set *Idc3* to zero and *Idc* to initiate a train of impulses in N2. Observe that this inhibitory synaptic interaction causes brief, positive current pulses in N1. These are outward currents that give rise to the negative membrane potential deflections (ipsp's) in N1. At high levels of firing in N2 (increase *Idc2*), the ipsp's sum to continuously hyperpolarize N1. Verify that these potentials are indeed inhibitory by first stopping impulse activity in N2 (turn the current off), then depolarizing N1 above threshold by driving the excitatory neuron N3 (by setting *Idc3*), and finally increasing *Idc2* until the inhibition from the N2-N1 synapse arrests the impulses in N1. Manipulate the impulse rates in N2 and N3 and observe that the spatial summation of the excitatory and inhibitory inputs sets the membrane potential, and therefore the impulse frequency, of the postsynaptic neuron, N1.

Synapse-5 Synaptic fatigue

For this exercise CIRCUIT is configured as in Synapse-3, with an inhibitory neuron N2 presynaptic to N1 (fig. I.6-4a). Unlike in the previous exercise, the synapse between N2 and N1 is set to undergo synaptic fatigue. Four windows are open when you begin this exercise (fig. I.6-6): two TIMESERIES windows, one that graphs the membrane potential (*Vm1*) of the postsynaptic neuron (N1) and another that graphs the presynaptic potential (*Vm2*); the PARAMETER window for controlling the potentials of N1 and N2; and the STIMULATOR window. The purposes of this exercise are to illustrate that synaptic potentials may vary with time and to explain how synaptic fatigue reduces the efficacy of

Figure I.6-6. Monitor display for exercise Synapse-5

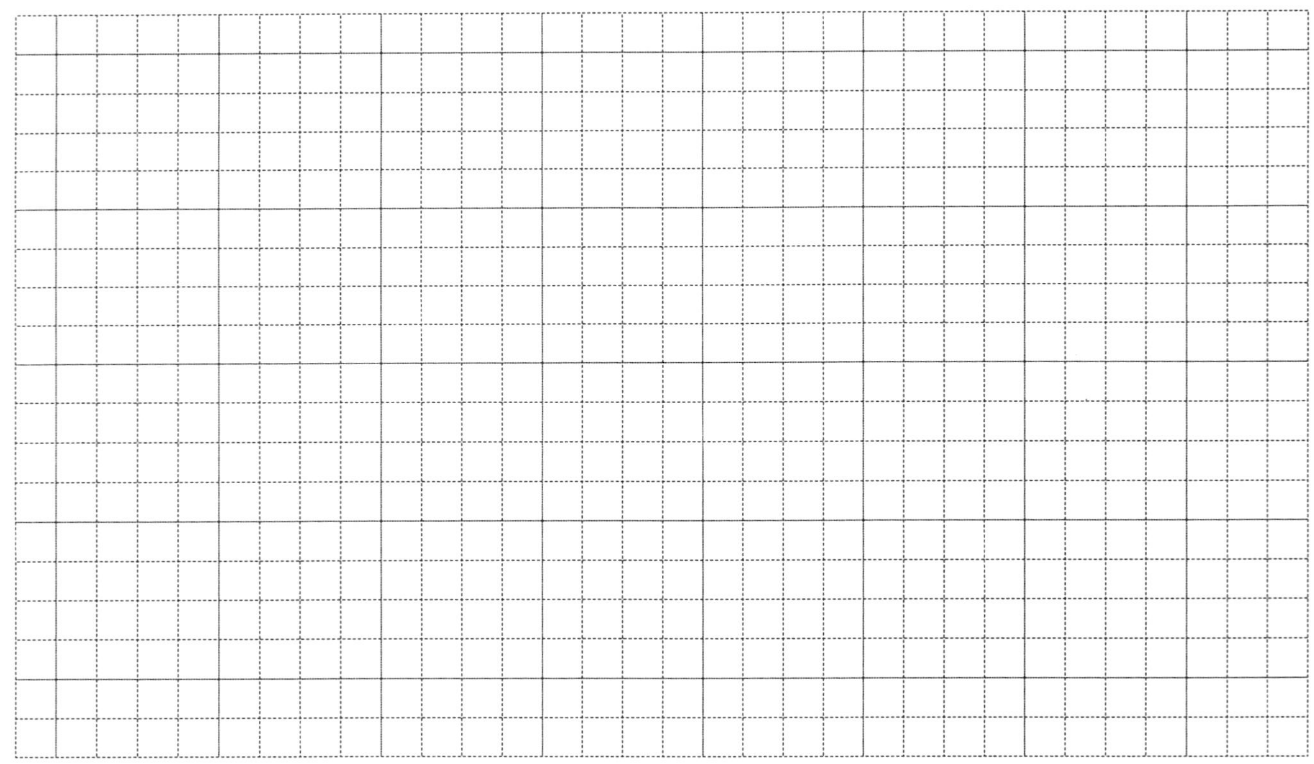

a strongly stimulated presynaptic neuron.

When this exercise begins (click on the STOP/GO icon), the postsynaptic cell (N1, upper window) is depolarized slightly (*Idc1* = 0.6 nA) to enhance the size of ipsp's. The presynaptic neuron is connected to the stimulator (*Stim2* = 1). Activate the stimulator to depolarize the presynaptic neuron (*Ampl* = 1.0 nA) and to generate a train of impulses. Notice that the first ipsp evoked in N1 is relatively large and that successive ipsp's are progressively smaller until a steady state occurs. The decrement occurs because the quantity of simulated transmitter in the presynaptic terminal is depleted by the spike train. Determine the rate of recovery from transmitter depletion by turning stopping the impulse train (turn the stimulator off), waiting about 0.3 s and then turning the stimulator on briefly to generate a single impulse. Measure the amplitude of this test ipsp and compare it with the amplitude of the initial, control ipsp. [Helpful hint: Use the ANALYSIS window, leaving the cursor in place but inactive between measurements.] Is recovery complete? Repeat this test by first obtaining a steady state of synaptic fatigue and then measuring the amplitude of a test ipsp for recovery times of 0.5 s, 0.7 s, and 1.0 s. Construct a graph of test ipsp amplitude versus the recovery interval, drawing a smooth curve between the points. Determine the time constant for recovery from synaptic fatigue from your graph.

After the synapse has fully recovered from synaptic fatigue, set the stimulator amplitude to 3 nA to generate a high frequency impulse train. You can see that the amplitude of the ipsp's evoked by these impulses quickly attenuate, becoming even smaller than those evoked by the lower frequency train. Because depletion is use-dependent, the higher level of firing engendered by the 3 nA current more nearly depletes the transmitter in the presynaptic terminal. When the presynaptic cell is driven by 4 nA, the synapse fatigues almost completely; now the membrane potential of N1 nearly returns to its resting level. To examine the functional consequences of synaptic fatigue, allow the synapse to recover (allow at least 5 s) and generate a slow train of impulses in N1 (set *Idc1* to 0.9 nA). Now excite N2 with the stimulator set to 4 nA. Note that initially the inhibitory synaptic interaction blocks impulse activity in N1; but as synaptic fatigue sets in, N1 begins generating impulses again at a frequency reduced only slightly from its original value.

Synapse-6 Non–spike-mediated synaptic interactions

For this exercise CIRCUIT again is configured as in Synapse-3, with an inhibitory neuron N2 presynaptic to N1 (fig. I.6-4a). For simulation there is no synaptic fatigue (simulating a very large supply of mobilized transmitter in the presynaptic terminal). Moreover, the impulse threshold for both neurons is set to a high value so that neither will generate impulses when depolarized; that is, both N1 and N2 simulate non-spiking neurons. The same four windows are open as in Synapse-3 (fig. I.6-7). The purpose of this exercise is to illustrate non–spike-mediated (also termed "chemotonic") synaptic transmission.

At the beginning of this exercise the postsynaptic cell (N1, upper window) is depolarized slightly (*Idc1* = 0.6 nA) to enhance the size of inhibitory inter

Figure I.6-7. Monitor display for exercise Synapse-6

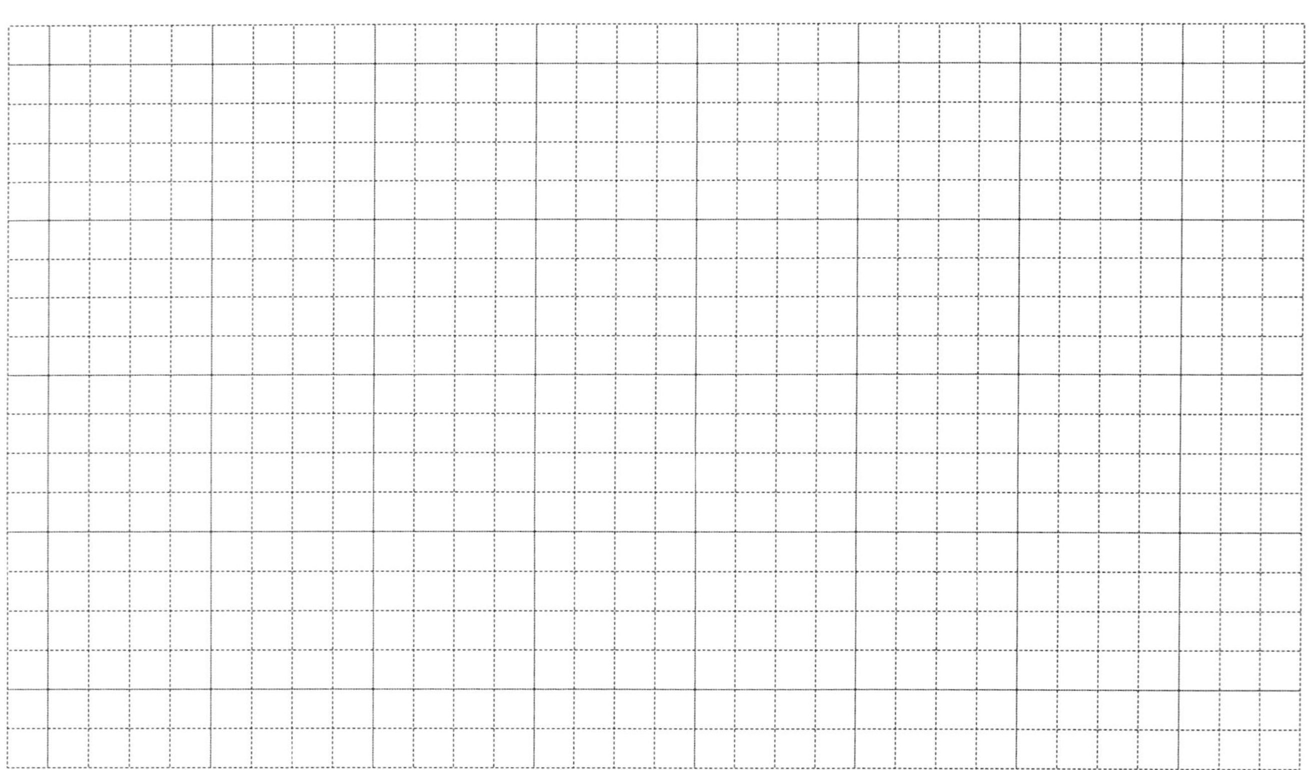

action between N2 and N1. The presynaptic neuron is connected to the stimulator (*Stim2* = 1). Activate the stimulator to inject depolarizing current pulses (1.0 nA) into the presynaptic neuron. Note that the depolarization induced in the presynaptic neuron by these current pulses exerts a hyperpolarizing effect on N1. The ipsp's differ from those generated by impulse primarily by their prolonged duration; in fact, the ipsp's last as long as the presynaptic depolarizations. Turn the stimulator off; then determine the presynaptic threshold for transmitter release by gradually depolarizing N2 with *Idc2* . Make a note of this value. Now turn on the stimulator again and observe that the N1 ipsp's are somewhat larger. Why? Leaving the stimulator to generate periodic depolarizing pulses, hyperpolarize the presynaptic neuron N2 in steps and examine the size of the N1 ipsp's. Not only do these decrease in amplitude, but there is now a significant synaptic delay between the onset of the presynaptic depolarization and the postsynaptic ipsp. Why? [Hint: Turn on the grid in the lower window and observe the relationship between the synaptic threshold and the membrane potential during the injection of the current pulses.]

MAXIMIZE the STIMULATOR window, reduce the duration of the current pulse to 10 ms, and set the amplitude to 10 nA. Then MINIMIZE the window and move it back to the lower right margin of the screen. Finally reset *Idc2* to zero. Now observe that the large impulse-like depolarizing pulses generated in N2 by the short-duration current pulses injected into N2 induce ipsp's in N1 that closely resemble nerve-impulse-evoked ipsp's. This result demonstrates that there is no fundamental difference between non–spike-mediated synaptic transmission and ordinary spike-mediated synaptic interactions between spiking neurons.

Synapse-7 Rectifying and non-rectifying electrical interactions

For this exercise CIRCUIT is configured for two neurons that are connected by simulated gap junctions. These may allow current flow in one direction only (rectifying) or in both directions (non-rectifying). Four windows (fig. I.6-8) are open when you begin this exercise: two TIMESERIES windows that graph the membrane potentials of N1 (*Vm1*, upper window) and of N2 (*VM2*, lower window) and two PARAMETER windows that are available for controlling the membrane potentials of the two neurons and for manipulating the strength of the interactions. The purpose of this exercise is to illustrate both rectifying and non-rectifying interactions and to show that rectifying electrical interactions can perform some of the same functions as chemical synaptic transmission. You will also see that non-rectifying interactions tend to equalize the impulse frequency in two neurons.

At the beginning of this exercise (click on the STOP/GO icon) the connection between N1 and N2 is non-rectifying; neuron N1 (*Vm1*, top graph) is depolarized slightly to generate nerve impulses at a moderate rate. The lower graph depicts the membrane potential of N2 (*Vm2*, bottom graph) at the same gain. Note that each impulse in N1 gives rise to a small excitatory potential in N2. This potential in N2 occurs without delay and reflects the impulse shape of N1 because it is caused by current flowing directly between these two cells.

Figure I.6-8. Monitor display for exercise Synapse-7

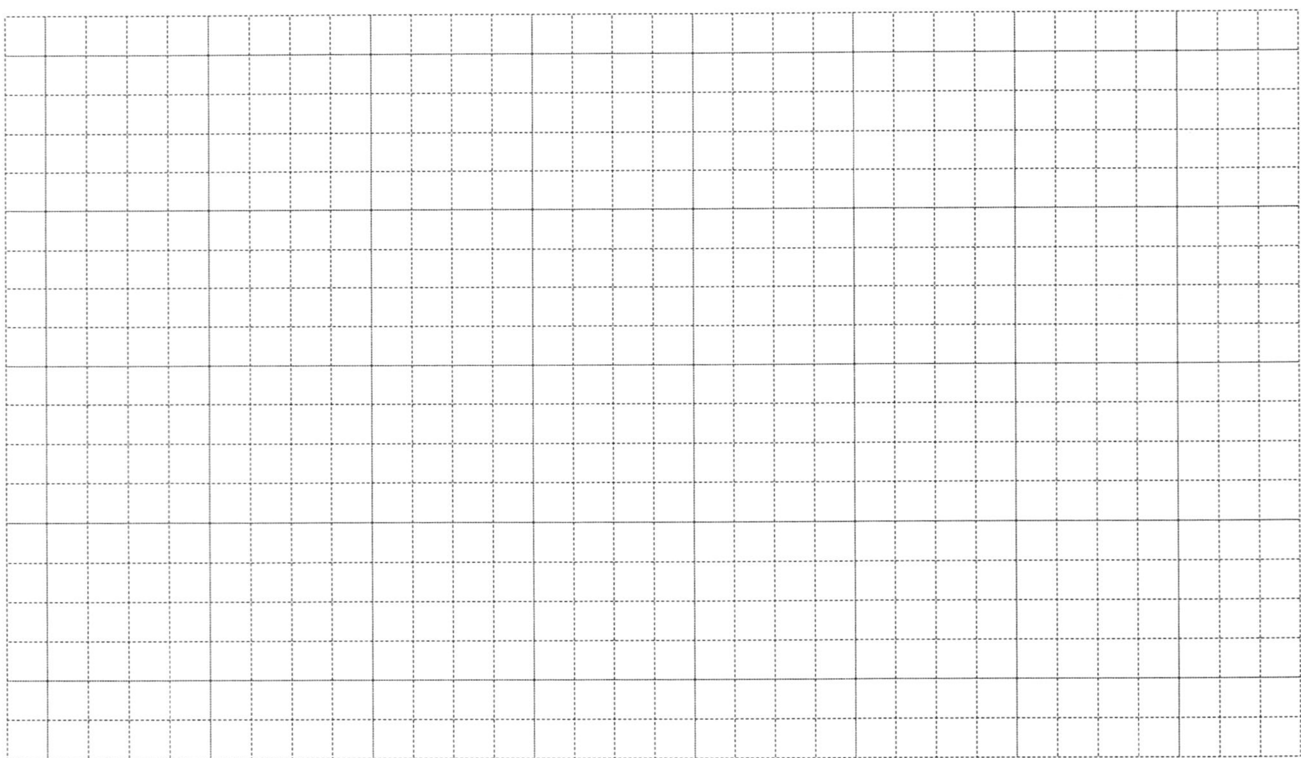

Open the appropriate PARAMETER MODIFICATION windows and increase the value of the parameters *gecp1* > *2* and *gecp2* > *1* (meaning "electrical coupling conductance" between N1 and N2 and between N2 and N1, respectively). This enhances the strength of the electrical coupling between these neurons. Note that this manipulation causes a greater response of N2 to N1 impulses. Increasing these parameters simulates increasing the conductance of the gap junction. By making this coupling strength sufficiently large (about 30 nS) the depolarizations induced by N1 impulses exceed impulse threshold in N2. At this point, each impulse in N1 evokes an impulse in N2; the two cells fire synchronously. Clearly, electrical interactions are not only fast but, as you see here, they also can be quite effective.

Apply a small hyperpolarizing current to N2 (about −0.4 nA) to block impulses in this neuron. You will see that this current spreads to N1 and reduces impulse frequency in this cell. Increase the amplitude of the current injected into N2 still further until (with *Stim2* about −0.6 nA) impulses are blocked in both cells. Thus sufficient inhibitory current has spread from N2 to N1 to indirectly block impulses in N1. To examine subthreshold electrical interactions, adjust the ordinate calibration so that the membrane potential ranges span about −75 to −55 mV in both graph windows and set *Idc1* to 0. Also set the coupling strength to be equal in both directions, with *gecp1* > *2* and *gecp2* > *1* both set to 20 nS. The membrane potentials of both cells should be at rest near −65 mV. Now hyperpolarize N1 by 10 mV and measure the change in the membrane potential in N2. The ratio of the change of the potential in N2 (this should be about 3 mV) to the 10 mV change in N1, the "coupling coefficient," is a measure of the strength of the electrical interaction. Verify that the coupling between these neurons is non-rectifying by showing that the coupling ratio again is about 0.3 if N2 is the neuron that is hyperpolarized. By repeating these procedures with depolarizing rather than hyperpolarizing current pulses, you will see that the sign of the change in membrane potential does not alter the measured coupling strength.

To examine the physiology of a rectifying interaction set *gecp2* > *1* to zero. The interactions are now set to mimic a rectifying electrical synapse in which current flows only from N1 to N2. Inject both depolarizing and hyperpolarizing currents into N1 and N2. Note the membrane potentials for each cell for all values of currents and make a table of your results. You will find that hyperpolarizing N1 does not alter the resting potential in N2; similarly, depolarizing N2 does not alter the resting potential in N1. The coupling coefficient apparently is zero; the cells are uncoupled. Tests of coupling strength in which N1 is depolarized or N2 is hyperpolarized yield the former value for the coupling coefficient, about 0.3.

Section I.7 Neuronal Oscillators

I.7.1 INTRODUCTION

In this section we address the question of how neurons form circuits that play specific functional roles in generating animal movements, in particular, rhythmic, repeating movements. We have known since the pioneering work of Adrian in the 1930s that the completely isolated nervous system can generate impulse patterns that are correctly structured for commanding meaningful movements. Specifically, Adrian showed that the respiratory rhythm in goldfish can be observed in the isolated goldfish medulla. However, not until after the research of Don Wilson on locust flight in the 1960s was it accepted that many of the *patterns* in animal movements are generated by the connections of central neurons (interneurons) in animal brains and spinal cords. The alternative view, that animal movements are purely sequences of reflexes requiring phasic sensory input, is no longer accepted.

Rhythmic (oscillatory) animal movements are characterized by their repetitious nature. Obvious examples include most locomotory movements such as walking, swimming, and flying and many visceral movements, such as breathing and the heartbeat. The neuronal activity that underlies rhythmic movements consists of bursts of impulses, separated by silent intervals. For all but the simplest movements several groups of neurons are active at different times in the movement cycle; hence, we speak of "multiphasic" activity. Such multiphasic activity is characterized by its repetition rate (frequency, or its inverse, period) and the phase relationships between activity in synergistic neurons. However the question remains, How do the nervous systems of animals generate the rhythmic impulse patterns that are translated by the muscles into these locomotory or visceral movements? There are several distinct models for the mechanisms by which neuronal circuits can give rise to oscillating patterns. We will address two general classes of models here, namely, recurrent cyclic inhibition (RCI) and reciprocal inhibition (RI). Both of these long-standing models have proven useful in enhancing our understanding of how neuronal circuits generate oscillations.

Many neuronal circuits with specific functions for rhythmic animals are now at least partially identified due to the intense interest in this topic during the 1970s and 1980s. These studies have been most successful in the invertebrates and some lower vertebrates, in which the relatively small number of nerve cells in the central nervous system allows individuals to be recognized by their anatomy and physiology. Although different neuronal circuits are associated with each type of animal movement, the interactions in these circuits have one important feature in common—most of the synapses are inhibitory.

I.7.2 MODELS FOR GENERATING OSCILLATIONS

I.7.2.1 Recurrent cyclic inhibition (RCI)

One model for how inhibitory neuronal circuits can generate oscillations, first proposed by Szèkeley, describes the activity of neurons connected to form a closed loop. The requirements for generating oscillations by this "recurrent cyclic inhibition" mechanism are simple. First, the loop must include an odd number of inhibitory neurons. Thus a loop of two cells must have one inhibitory and one excitatory neuron, whereas a loop of three cells could function either with three inhibitory neurons or with one inhibitory and two excitatory neurons. Second, the inhibitory interactions in the loop must be strong enough so that the depolarization of the presynaptic neuron suppresses activity in the postsynaptic neuron. Third, sufficient excitation must be supplied to all neurons in the loop to ensure that neurons will generate impulses when released from inhibition. An RCI circuit formed by three inhibitory neurons (assumed to be spontaneously active without any external drive) is shown in figure I.7-1. In this circuit, each neuron is sequentially active, then inhibited, and finally recovering from inhibition. The period of the cycle is regulated by the activity level of the neurons. At higher levels of excitation the recovery from inhibition occurs more quickly, and hence the cycle period is briefer. Neurons in this circuit are active (generate impulses) sequentially, with phase lags of 120°; the progress of the activity around the loop is in the reverse direction of the inhibitory connections (see modeling exercise Circuit-1).

I.7.2.2 Reciprocal inhibition (RI)

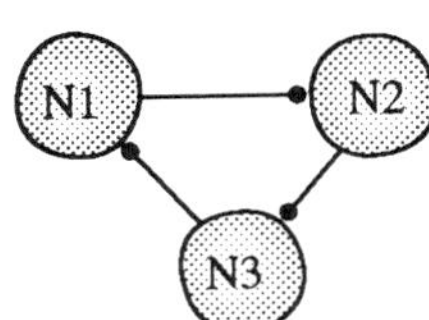

Figure I.7-1. Circuit diagram for recurrent cyclic inhibition (RCI)

Although the mechanism of RCI described above will not generate oscillations in even-membered loops, such as in a pair of mutually inhibitory neurons, a mechanism called "reciprocal inhibition" will generate oscillations in this simple, even-membered circuit. The neuronal circuit envisioned by the RI model is that of two mutually inhibitory neurons which are driven by excitatory input from a third neuron (fig. I.7-2a) or by current injection into both cells. This circuit will generate oscillations once the excitatory neuron is activated provided that the inhibitory neurons or their synaptic interactions have some dynamic, special properties that function to limit the duration of the inhibition exerted by one neuron on the other. Three such neuronal properties are 1) synaptic fatigue,

2) postinhibitory rebound (PIR), and 3) some form of delayed excitation. The output of this circuit consists of impulse bursts alternating with inhibition in each of the neurons. The membrane potential oscillations and impulse bursts of the two inhibitory neurons are antiphasic (the cells oscillate 180° out of phase).

I.7.2.3 Synaptic fatigue

Circuits in which mutually inhibitory neurons undergo synaptic fatigue to limit the duration of synaptic inhibition (perhaps because of transmitter depletion or inactivation of the calcium influx in the presynaptic terminals) are found in several identified neuronal oscillator circuits. We can understand how an RI circuit with synaptic fatigue can generate oscillations by the following considerations (figs. I.7-2a,b). Suppose that the excitatory neuron in this simple circuit is suddenly activated. Either inhibitory cell I1 or cell I2 will become active first (because of inevitable asymmetries). Let's assume that cell I1 generates impulses first and inhibits cell I2. This inhibition keeps cell I2 off only for a limited time because the synaptic fatigue reduces the efficacy of the inhibitory synapse. When the inhibition has decreased sufficiently, cell I2 becomes active because of the continuous excitatory drive. When cell I2 exceeds threshold, it inhibits cell I1, thereby removing the last remanent of inhibition it receives from cell I1. Two processes occur while cell I2 is active. First, the synapse of cell I1 with cell I2 recovers from synaptic fatigue; for example, more neurotransmitter may be mobilized. Second, the synapse from cell I2 to cell I1 undergoes synaptic fatigue and becomes less effective in depressing the activity of cell I1. Because of the onset of fatigue in the cell I2–cell I1 synapse, cell I1 resumes generating impulses and depresses the impulse activity in cell I2. This step completes one cycle. Obviously, this model system will only generate alternating bursting oscillations if the drive from the excitatory neuron is sufficiently great to drive both cells and if the excitation supplied by this neuron is not too great to be countered by the synaptic inhibition (see modeling exercise Circuit-2).

I.7.2.4 Postinhibitory rebound (PIR)

A second cellular property that can aid in generating oscillations in reciprocally inhibitory neuronal pairs is "postinhibitory rebound." PIR (also called "paradoxical excitation" and "anodal break excitation") describes the overshooting, excitatory response of neurons at the abrupt termination of inhibitory input. That is, in many neurons termination of an imposed inhibition, whether by passing current or through synaptic inhibition, is followed by a brief depolarization and an increased impulse activity. One identified source of this excitation is the activation of an inward, excitatory current by hyperpolarization, I_h, for example. Another possibility is the de-inactivation of a slow sodium current when the membrane potential of a cell is hyperpolarized. For this mechanism to generate oscillations, no external excitatory input is required. Assume that the two inhibitory neurons in figures I.7-2a,c exhibit PIR; if cell I1 is strongly hyperpolarized and then released, it will respond with a transient depolarization and a burst of impulses. This excitation induces inhibition in cell

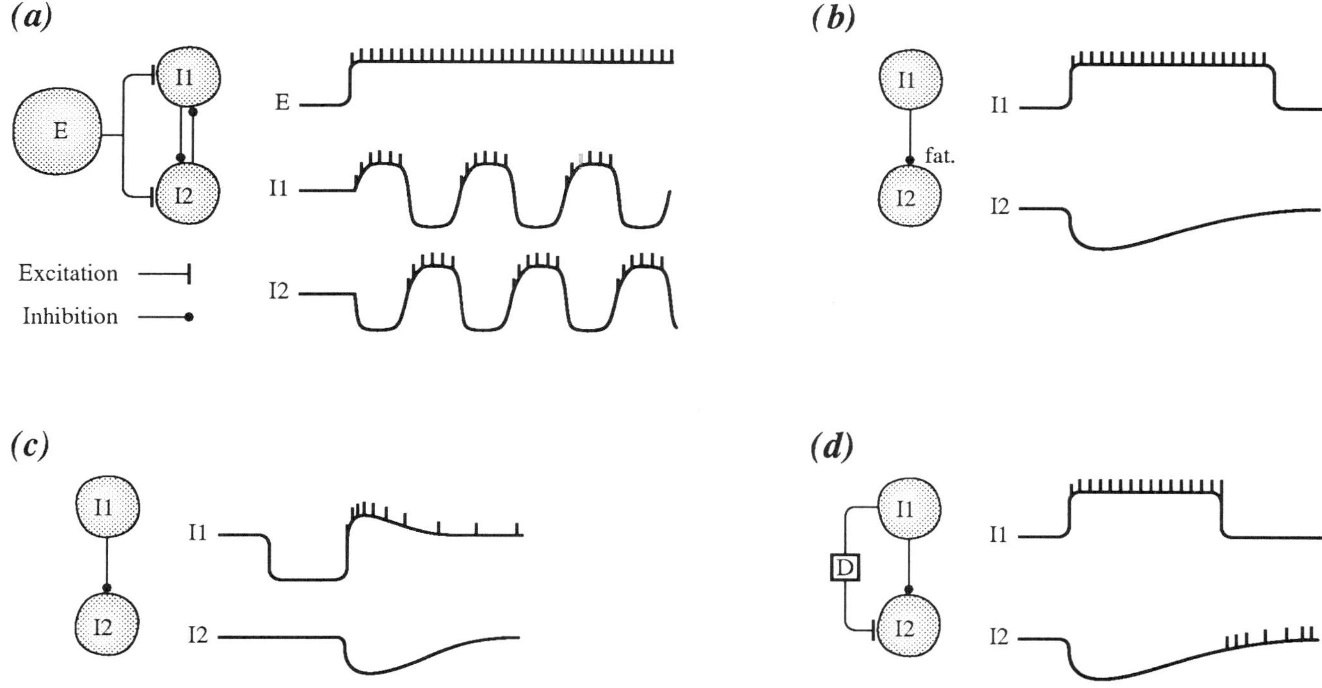

Figure I.7-2. Circuit oscillator with reciprocal inhibition (RI). *a*) Schematic diagram of a circuit generating oscillations via RI. *b*) Fatiguing synaptic interaction. *c*) Circuit with postinhibitory rebound (PIR). *d*) Circuit with delayed excitation

I2 but only for a short interval because PIR is short-lived. On being released from inhibition, cell I2 also may undergo PIR and inhibit cell I1. Thus reciprocal bursting in the two cells could persist indefinitely without external input except for the initial kick to one neuron (which may be excitatory or inhibitory) to start the oscillations. Such bursting will be sustained only if the inhibition and the PIR are strong (see modeling exercise Circuit-3).

I.7.2.5 Delayed excitation

Delayed excitation is a third means that, when found in a circuit of reciprocally inhibitory neurons, can function to generate neuronal oscillations. The idea here is that such a circuit of two inhibitory neurons will oscillate in antiphase if the inhibition exerted by one neuron on the second is followed and superseded by excitation. The excitation could arise from a second, slower excitatory synapse that is connected in parallel with the inhibitory one. Such dual synapses are found in a marine slug. Refer to figure I.7-2d to understand how RI circuits with delayed excitation can generate oscillations. Assume that in this circuit cell I1 receives strong external excitatory drive (not shown). The cell I1–cell I2 synapse hyperpolarizes cell I2 and depresses activity in this cell. With some delay, *D* in the diagram, cell I2 hyperpolarization is reversed by the second, slow excitatory synapse (fig. I.7-2d). Hence, cell I2 begins to depolarize and escapes from cell I1 inhibition to generate nerve impulses. If these cells were connected by reciprocal inhibition, cell I2 activity would now begin to inhibit cell I1. This is the principle underlying the swim oscillations in the marine sea slug *Tritonia*.

I.7.3 RECIPROCAL INHIBITION GENERATES THE HEART-TUBE OSCILLATIONS IN *HIRUDO MEDICINALIS*

One example of a neuronal system that generates oscillations via the mechanism of RI is the rhythmic contraction of heart tubes in the medicinal leech, *Hirudo medicinalis*. The leech heart consists of bilateral tubes whose peristaltic contractions propel the blood in the two lateral blood vessels. The movement pattern in the heart tubes is the result of alternating contractions and relaxations in the circular muscles. As in other cardiac systems these heart tubes are active continuously; in leeches the cycle period is about 15 s. Much of what we know about the function of the leech cardiac system is the result of research performed by Wes Thompson, Gunther Stent, Ron Calabrese, and their colleagues.

The neuronal control of this ongoing visceral movement includes at least two levels of organization. First, there are heart interneurons, whose special properties and specific interactions actually generate the oscillation. These inhibitory neurons occur in the anterior segments of the leech ventral nerve cord. Second, heart motor neurons, which are repeated in many segments of the leech ventral nerve cord, receive oscillatory input from the heart interneurons and in turn control the contractions of the circular muscles of the heart tubes in this neurogenic system. Here we consider only the antiphasic aspect of the activity in the oscillator interneurons and not the conversion of interneuron activity into the more complex oscillations found in heart motor neurons.

The entire oscillator circuit includes several sets of reciprocal inhibitory interactions between the interneurons (fig. I.7-3). One peculiarity of this circuit is that, as for many other interactions in the leech, the interactions are mediated both by impulses and by chemotonic means. The interneurons receive only weak external inputs; nevertheless, the interneurons all generate nerve impulse trains (but generate no bursting oscillations) when the inhibitory synapses are blocked. The neurons and synaptic connections in this circuit exhibit several dynamic properties that ensure the occurrence of oscillations and set the cycle period. Three of the self-limiting neuronal properties described above are found in the heart interneurons: 1) postinhibitory rebound; 2) delayed excitation, caused by hyperpolarization-activated inward cation current (the I_h current described in section I.5); and 3) synaptic fatigue. The presence of any one of these three properties could, at least in theory, generate the cardiac oscillations in the leech heart circuit. Taken together, they ensure that the heart interneurons form a reliable oscillator to generate impulse bursts that time the contraction of the leech heart tubes (see modeling exercise Circuit-4).

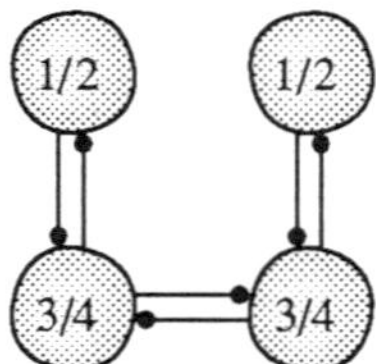

Figure I.7-3. Neuronal circuit generating the *Hirudo medicinalis* cardiac oscillations. Circles designate left and right homologs of heart interneurons (HN) in leech midbody segments 1 and 2 (1/2) and 3 and 4 (3/4). The lines ending in filled circles show the reciprocal inhibitory connections between HN cells in these segments.

I.7.4 RECIPROCAL INHIBITION TOGETHER WITH DELAYED EXCITATION GENERATES THE SWIM OSCILLATIONS IN *TRITONIA*

As a second example of how oscillations are generated in an identified neuronal circuit, we will consider the network that generates the escape behavior of *Tritonia*, a marine slug that resembles an overgrown terrestrial banana slug. Most of our understanding of how this system functions was uncovered during the last twenty years in the laboratories of A. O. D. Willows and Peter Getting. *Tritonia* are preyed upon by starfish. To escape this predator, which they detect when contacted by the starfish tube feet, *Tritonia* initiate a swimming movement that consists of about five to ten cycles of alternating contractions in dorsal and ventral muscles generating, respectively, dorsal and ventral flexions. The activity is characterized by two phases that are 180° out of phase with each other (in antiphase). The cycle period of each dorsal and ventral flexion is about 5 s, increasing gradually to about 10 s as the swim episode progresses.

The neuronal control of this escape behavior includes at least three organizational levels: sensory neurons, which are activated by the stimulus from the starfish; interneurons, which receive excitatory input from the sensory neurons and whose interconnections generate the two-phase oscillations that underlie the swimming behavior; and motor neurons, which receive excitatory drive from the interneurons and pass this information onward by commanding the appropriately timed contractions of the dorsal and ventral muscles of the slug. Many details concerning the functioning of the sensory level of control remain to be discovered. We do know, however, that in response to appropriate stimulation the system of about 80 sensory neurons conveys prolonged (but slowly decaying) excitatory input to the interneurons that form the central oscillator. It is this "ramp excitation" that not only initiates the neuronal oscillations (and consequently the movements of the animal) but also ends the oscillations and the swimming as the level of excitation decreases below threshold levels. The motor neurons appear to be largely follower cells that relay information on to the muscles, so we will ignore them in the following discussion of how the interneurons generate the cycle period and phase relationships of the swim oscillations.

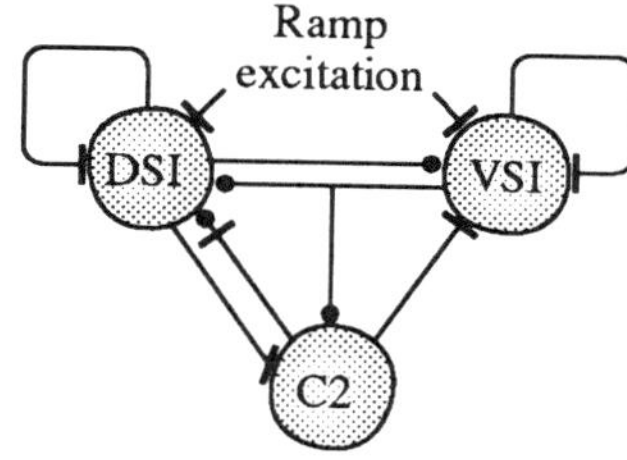

Figure I.7-4. Neuronal circuit generating swimming in *Tritonia*. The DSIs (dorsal swim interneurons) are connected via reciprocal inhibition to the VSIs (ventral swim interneurons). In addition, DSIs have excitatory synapses with the C2 neurons, the VSIs inhibit the C2s, and both DSIs and VSIs have self-excitatory synapses. The C2s are connected by a mixed excitatory and inhibitory synapse to the DSIs and, via an inhibitory synapse, to the VSIs. Excitation is supplied to this system by sensory input shown as "Ramp excitation" in this illustration. This input is strongest for the DSI neurons.

The circuit of interneurons that generates the swim oscillations in *Tritonia* includes three types of neurons (fig. I.7-4). There are six Dorsal Swim Interneurons (DSIs) that are excited from synaptic contacts from the sensory cells. There are also self-excitatory interactions among the DSIs, which act to enhance the excitatory input from the sensory cells. The DSIs provide excitatory synaptic drive to two interneurons named "C2." The two C2s are neurons with high thresholds; that is, they require prolonged, intensive stimulation before they generate impulses of their own. Thus there is an appreciable delay between the beginning of impulse activity in the DSIs and the C2s. The DSIs also have a second set of synaptic outputs; they make inhibitory synapses with the four Ventral Swim Interneurons (VSIs). These VSIs also receive some excitatory input from sensory stimulation. Other circuit interactions are the excitatory connections from the C2 neurons to the VSIs and the synapses by which the VSIs inhibit both the DSIs and the C2s (fig. I.7-4). Thus the connections in this oscillatory circuit include several excitatory connections in addition to reciprocal inhibition between the DSIs and VSIs. Although not explicitly included in figure I.7-4, many of the interconnections between DSIs, VSIs, and C2s are complex, consisting of multimodal synapses that include both fast- and slow-acting conductances, which are both inhibitory and excitatory.

Getting has proposed that the *Tritonia* swim circuit should be viewed as an example of a neuronal oscillator that functions because of reciprocal inhibition combined with delayed excitation. To understand this model, assume that the sensory cells are activated by some external stimulus. They drive the DSIs by prolonged, ramp depolarization. The excited DSIs inhibit the VSIs and keep them suppressed while simultaneously exciting the C2s. The C2s respond to the DSI input by slow depolarization; they are not inhibited by the VSIs because these are shut off by the DSIs. After some interval, the C2s generate impulses that provide excitatory input to the VSIs, countering the inhibition from the DSIs. If this excitation is sufficiently strong, the VSI membrane potential will rise above threshold, and these cells will begin to generate nerve impulses. There are two consequences of VSI activity. The DSIs are inhibited, thereby removing the excitation from the C2s. Also, the C2s are inhibited directly, but slowly, by the VSIs. With the excitatory level in C2s reclining, the excitatory drive to the VSIs also declines. Eventually they fail to keep the DSIs inhibited in the face of continuing excitatory input to the DSIs from the sensory neurons. The DSIs, released from the VSI inhibition, begin to generate impulses and hence initiate a new cycle. These cycles repeat until the excitation from the sensory cells to the DSIs wanes sufficiently so that the DSIs no longer generate impulse activity when the VSI inhibition is absent (see modeling exercise Circuit-4).

Modeling Exercises

***NeuroDynamix* Model: CIRCUIT   Exercises: Circuit-1 to Circuit-4**

Introduction

This series of modeling exercises illustrates how oscillations are generated by some simple neuronal circuits. Two types of circuits are demonstrated, recurrent cyclic inhibition in a ring of three neurons and three circuits with reciprocal inhibition between pairs of neurons. Dynamic cellular and circuit processes demonstrated by these exercises include synaptic fatigue, the h current, and delayed excitation provided by a third, excitatory neuron. To carry out these exercises you will be using the same model, CIRCUIT, that generated the Synapse exercises. After you have completed each of the exercises outlined here, feel free to alter model parameters, such as synaptic amplitude and time course, to get a better feel for how neurons interact. Use the interactive FIGURE icon to view the circuit simulated by each of these exercises.

Circuit-1 Recurrent cyclic inhibition

For this exercise CIRCUIT is configured with three inhibitory neurons that are synaptically interconnected to form an inhibitory loop. In this simulation, all three neurons are configured without any time-dependent currents other than the currents generated by the inhibitory synaptic interactions. The inhibitory loop is formed because in this circuit neuron 1 (N1) inhibits neuron 2 (N2), N2 inhibits neuron 3 (N3), and N3 inhibits N1 to close the loop (fig. I.7-1). To illustrate the output of this loop of recurrent cyclic inhibition (RCI) three TIMESERIES windows are open, one for the membrane potential of each neuron. These windows are arranged to suggest the interaction loop with *Vm1* graphed at the upper left, *Vm2* graphed at the upper right, and *Vm3* graphed directly below the other two (fig. 1.7-5). Two additional windows are open, the PARAMETER window to control synaptic parameters and the STIMULATOR window to control the excitatory drive to the three neurons. The purpose of this exercise is to illustrate how a simple neuronal circuit that comprises three inhibitory neurons can generate stable oscillations.

When this exercise begins (click on the STOP/GO icon), the three neurons in this circuit are quiescent, below impulse threshold. The membrane potentials of all three neurons in the RCI circuit are controlled by the stimulator; each one receives the same amount of current when the stimulator is turned on. In the absence of current injection there clearly are no oscillations. Now set the stimulator amplitude to 1 nA (*Ampl* = 1). Notice that all three neurons are depolarized by this current but that one of them attains threshold first and generates a neuron impulse. (This occurs at random because of the random membrane potential noise associated independently with each neuron.) After some initial, apparently disorganized impulse activity, a regular pattern of impulse activity emerges. Study this pattern carefully. Note that the neurons are in three different states, which are passed sequentially from one cell to the next.

Figure I.7-5. Monitor display for exercise Circuit-1

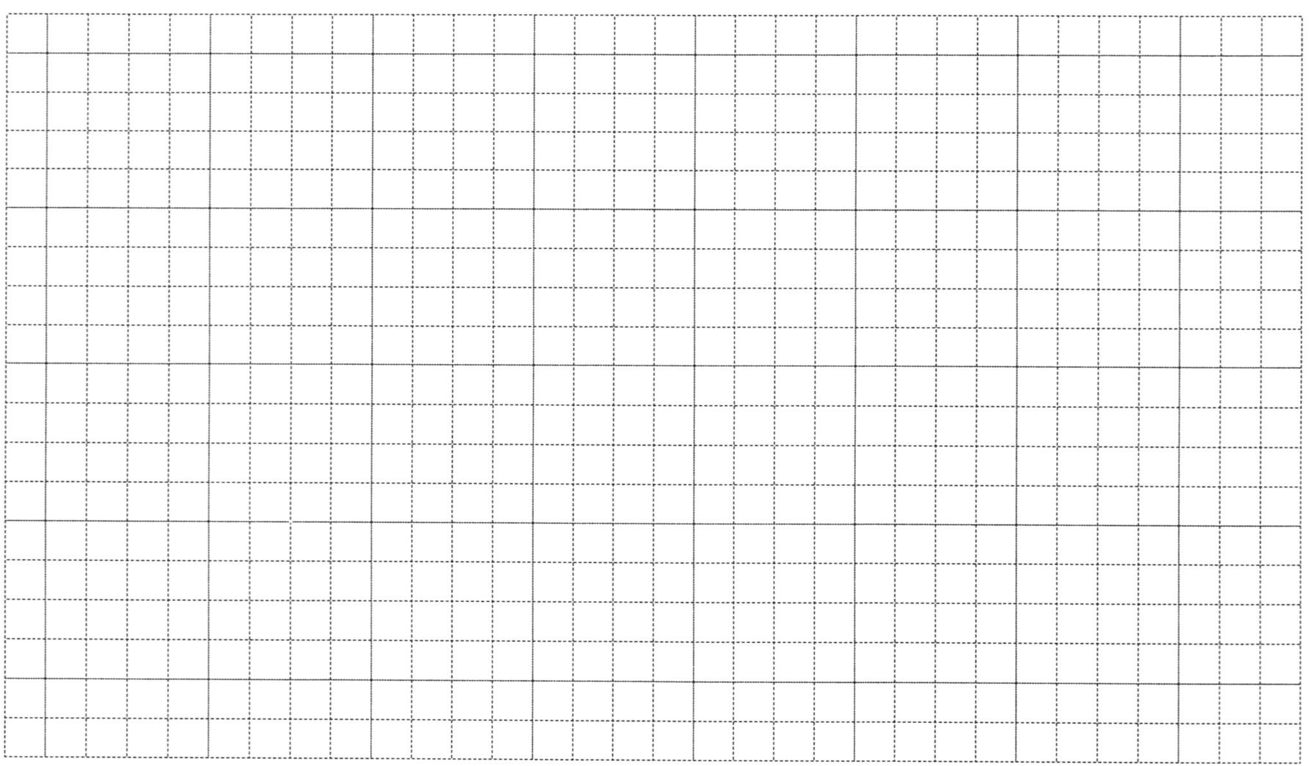

When one cell is generating impulses, a second one (its postsynaptic target clockwise in the loop) is inhibited by these impulses, while the third one is recovering from previous inhibition. You will see that these states progress *counterclockwise* around the loop, in the opposite direction of the inhibitory interactions.

The period of the oscillations in this simple RCI loop is the sum of the durations of the three states, each of which lasts for 1/3 of the cycle period. In fact, the cycle period is set by the sum of the durations of the recovery states for each of the neurons. Verify this statement by stopping the graphing when the traces reach the right side of the graph (click on the single arrow icon in one of the TIMESERIES windows) and measuring the duration of the recovery state. Note also that the impulse activity of the three neurons occurs in three phases, each separated by 1/3 of the cycle (120°).

Neuronal circuit oscillators within the central nervous system underlie the rhythmic movements expressed during animal locomotion. Animals control the speed of locomotion by changing the period of the underlying neuronal oscillators. By increasing the excitation applied to the neurons in this simulation via the current amplitude of the stimulator, you can decrease the cycle period (increase the frequency) of the oscillations in this RCI circuit. To examine the relationship between applied excitation and cycle period, set *Ampl* to a series of increasingly larger values (2, 3, and 4 nA). Notice that the period decreases with each increase in the stimulus amplitude. The decrease in period is caused by the decrease in the length of the recovery state on each cycle; cycle period is still the sum of the duration of the recovery states. Make a graph of cycle period versus the amplitude of the excitatory current.

The only time-dependent property required for RCI loops to generate oscillations is that recovery from inhibition not be instantaneous. There is a further requirement, however; the number of inhibitory interactions in the loop must be odd (three in the case we have studied). Explore this requirement by reconfiguring the circuit for an even number of inhibitory neurons (2 and 4). Does the circuit oscillate? You should observe that two states are possible with these even-numbered loops. Both of these states are stable; hence, no oscillations can occur. You may also explore other odd-numbered loops by forming circuits with two excitatory and one inhibitory synapse or one excitatory and three inhibitory synapses. To generate these additional circuits, set the synaptic parameters *Ain* and *Aex* appropriately. You will also need to set the parameter *NumCells* to 4 when simulating circuits of four neurons.

Circuit-2 Reciprocal inhibition oscillator: synaptic fatigue

For this exercise CIRCUIT is configured to simulate a two-neuron inhibitory loop (fig. I.7-5a). As we learned in the previous exercise, such even-numbered inhibitory loops will not generate oscillations in neurons without time-dependent conductances. The inhibitory loops explored in this and in the succeeding exercises *do* oscillate because the neurons or the circuits formed by the neurons include dynamic (time-dependent) processes. For this exercise the inhibitory interactions are time dependent; namely, the strength of the inter-

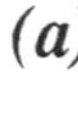

(a)

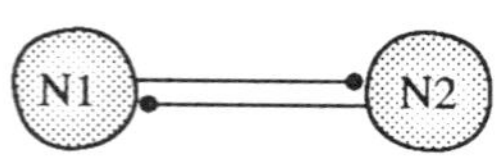

Figure I.7-6. Reciprocal inhibition. *a*) Circuit diagram for RI. *b*) Monitor display for Circuit-2

(b)

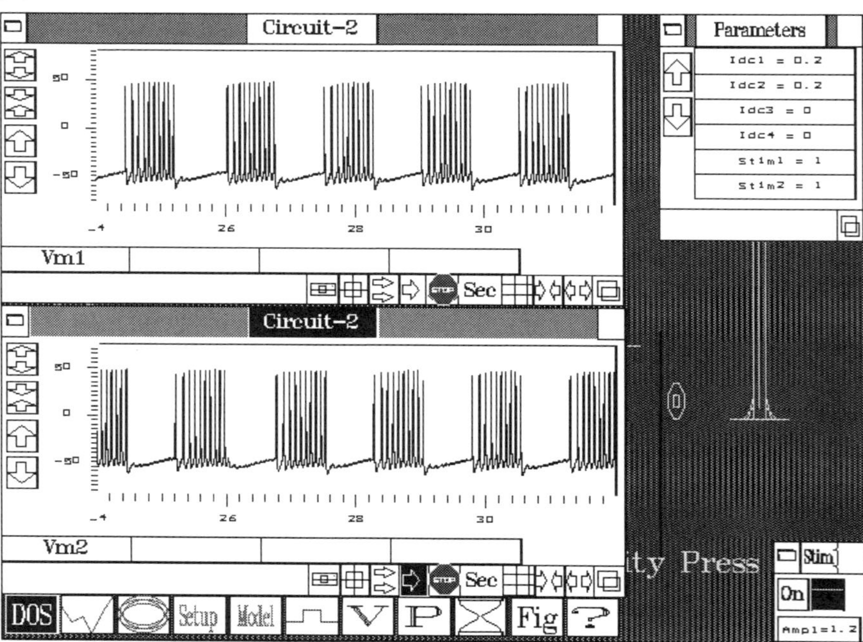

action decreases as a function of time. Two TIMESERIES windows are open for this exercise, one for the membrane potential of each neuron (N1, upper; N2, lower; fig. I.7-6b). Two additional windows are open, the PARAMETER window to control synaptic parameters and the STIMULATOR window to control the excitatory drive to the neurons. The purpose of this exercise is to illustrate how oscillations can be generated by a two-neuron, reciprocally inhibitory loop if the synaptic interactions undergo synaptic fatigue.

Begin the exercise by clicking on the STOP/GO icon; at the initial parameter settings the two neurons are quiescent. [Hint: You may wish to set the *skip* factor to 3 or 5 (use the CLOCK icon) in order to increase the graphing rate.] Drive N1 with excitatory current (set *Stim1* to 1, and turn the stimulator on with *Ampl* set to 1.2 nA) to generate a high frequency impulse rate in N1. Because of the inhibitory synapse from N1 to N2, the excitation of N1 sharply hyperpolarizes N2. Observe that this inhibitory effect decreases gradually so that after a second or two the membrane potential has returned to a less hyperpolarized steady-state level. Clearly the synapse from N1 to N2 undergoes synaptic fatigue. Now provide excitatory drive to N2 by connecting it to the stimulator as well (set *Stim2* to 1). This excitatory current is sufficient to bring the membrane potential of N2 above impulse threshold (even though the inhibition from N1 is still present). Because the N2–N1 synapse has not been active (N2 had not been generating impulses), the strong inhibition provided by N2 overcomes the excitatory drive from the stimulator and sharply inhibits N1. This inhibitory synapse also fatigues, and the potential in N1 becomes less hyperpolarized until it crosses threshold and N1 resumes impulse activity. Meanwhile the N1–N2 synapse has been recovering from its fatigued state and again strongly inhibits N2. As you can observe, neither cell remains active for long; instead, each alternates between an excited and an inhibited state.

Observe first the relationship between the impulses in N1 and N2. What is the cycle period of the oscillation in each neuron? It should be about 1.5 s. Note that the two neurons are active in antiphase. When one is firing impulses, the other is inhibited. Antiphasic activity cycles are characteristic of neurons connected by reciprocal inhibition. Oscillations in this circuit are generated only when the excitation level is in a very limited range. Explore this range by first increasing the amplitude of the stimulator current for both cells to 2.2 nA. At this level of stimulation the cells exhibit continuous impulse trains rather than alternating bursts of impulses. The impulses may occur synchronously in the two cells or they may alternate; either of these conditions can be stable for an RI oscillator with excessive excitation. Now return the stimulus current to 1.2 nA and wait until stable oscillations have resumed. Note the cycle period at this level of stimulation. Reduce the excitatory current to both N1 and N2 further (by setting *Idc1* and *Idc2* to lower, even negative values) until there are no antiphasic oscillations. Note the minimum current required to sustain oscillations and the cycle period at this level of stimulation. Increase the current to both neurons in steps until the excitatory drive is too great to allow for bursting oscillations. Note the maximum current that can support bursting oscillations and the cycle period at this stimulus level. The minimum and maximum current values define the stability limits of this circuit for excitatory drive. As you can see, this limit is much narrower than for the RCI oscillator of the previous exercise.

Figure I.7-7. Monitor display for exercise Circuit-3

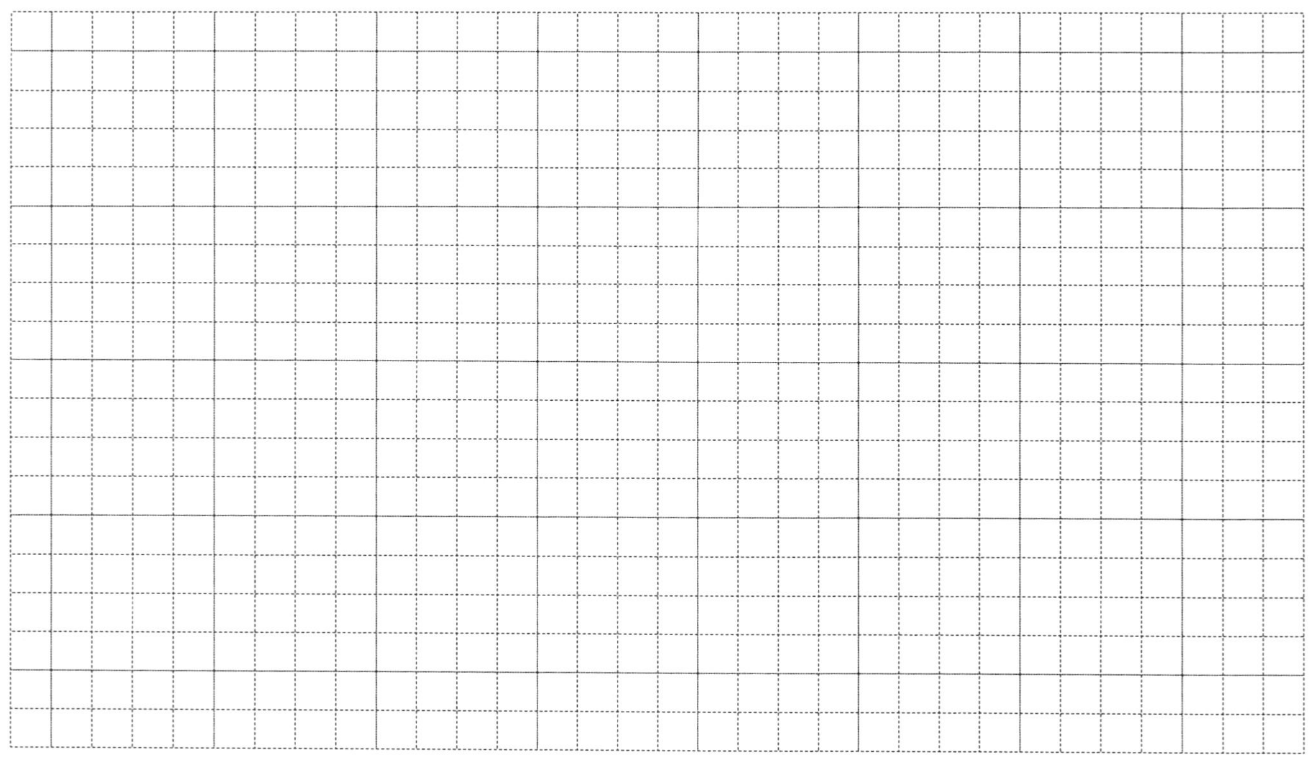

Circuit-3 Reciprocal inhibition oscillator: h current and postinhibitory rebound

For this exercise CIRCUIT again is configured to simulate a two-neuron inhibitory loop (fig. I.7-6a). The synapses in this exercise are non-fatiguing because the aim here is to explore oscillations generated when each neuron includes the h conductance, which is activated on hyperpolarization. The circuit is meant to simulate the heart oscillator in the medicinal leech. Recall from exercise Neuron-5 that the h conductance acts to counter hyperpolarization by generating an inward, excitatory current. The activation kinetics of the h conductance underlie the dynamics of the oscillations generated by this simple RI circuit. Two TIMESERIES windows are open for this exercise, one for the membrane potential of each neuron (N1, upper; N2, lower; fig. I.7-7). Two additional windows are open—the PARAMETER window to control synaptic parameters and the STIMULATOR window to control the excitatory drive to the neurons. The purpose of this exercise is to illustrate that oscillations can be generated by a two-neuron, reciprocally inhibitory loop if the neurons individually have special dynamic properties due to the h conductance.

Begin the exercise by clicking on the STOP/GO icon. [Hint: You may wish to set the *skip* factor to 3 or 5 (use the CLOCK icon) in order to increase the graphing rate.] Notice that even though neither cell is receiving any external excitatory drive, both are generating impulses; i.e., their membrane potential as the graphing begins. Soon one cell dominates; its impulse rate increases while that of the other neuron decreases and soon ceases. However this is not a stable state. After firing for a short interval, the second neuron becomes active inhibiting the first. This alternation of activity, oscillations of the two neurons in antiphase, continues indefinitely. No external excitation is required because the h current, activated each time a neuron is inhibited, provides the necessary excitatory drive.

Measure and record the cycle period of the unstimulated oscillator circuit. Now investigate the dependence of cycle period on external excitatory drive by applying a current of 1 nA to each neuron (set stimulator *Ampl* to 1). Notice that the increased excitation causes and increase, rather than a decrease, in cycle period, in contrast to our observations on the RCI circuit and the RI circuit with synaptic fatigue. In fact, length of period expressed by the oscillator with this level of external excitation, about 12 s, approximates the period observed in the leech heart oscillator. The oscillations in this circuit are very robust. Verify that they persist even when the 1 nA hyperpolarizing current is injected into the two neurons. Note that the cycle period of the oscillator now is short.

Turn off the stimulator and scroll down the list in the PARAMETER window to find the parameter *ghMax*, which sets the maximum value of the h conductance. Open the PARAMETER MODIFICATION window for this parameter and change its value from 100 nS to 0. The absence of impulses in the two neurons with *ghMax* set to zero demonstrates that the excitation in this oscillator is generated internally by the h current. On resetting the maximum h conductance to 100 nS, the oscillator immediately resumes its cycling. For further investigation of the dependence of cycle period on model parameters, reduce the values of the activation time constants for the h conductance

(a)

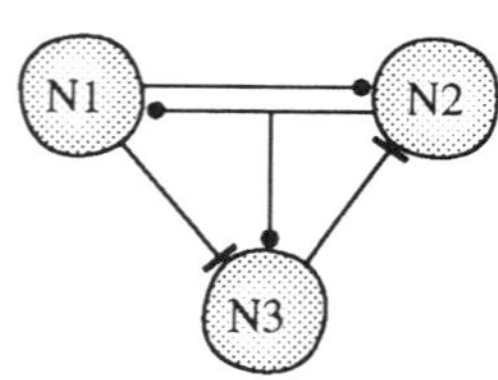

(b)

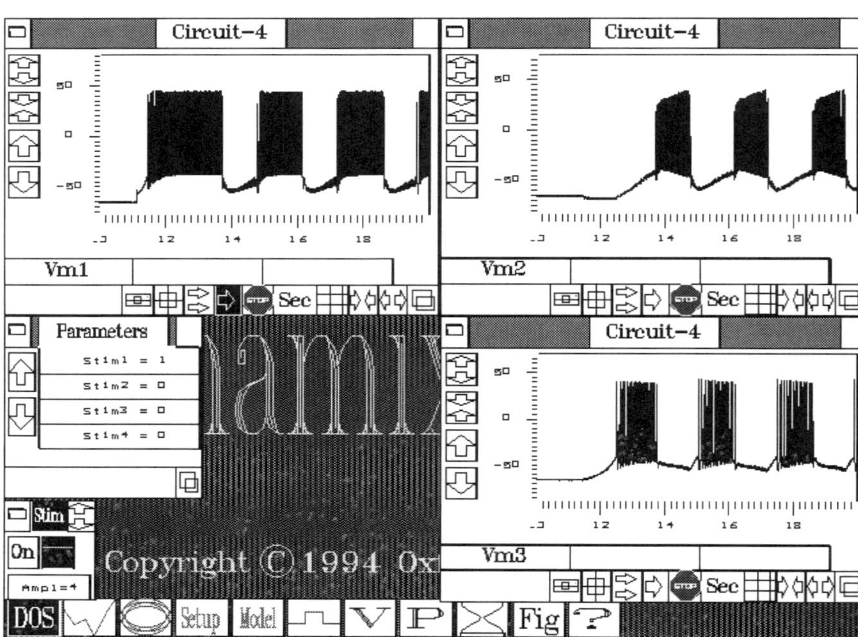

Figure I.7-8. *Tritonia* swim circuit model. *a*) Simplified circuit diagram for the swim circuit in *Tritonia*. *b*) Monitor display for Circuit-4

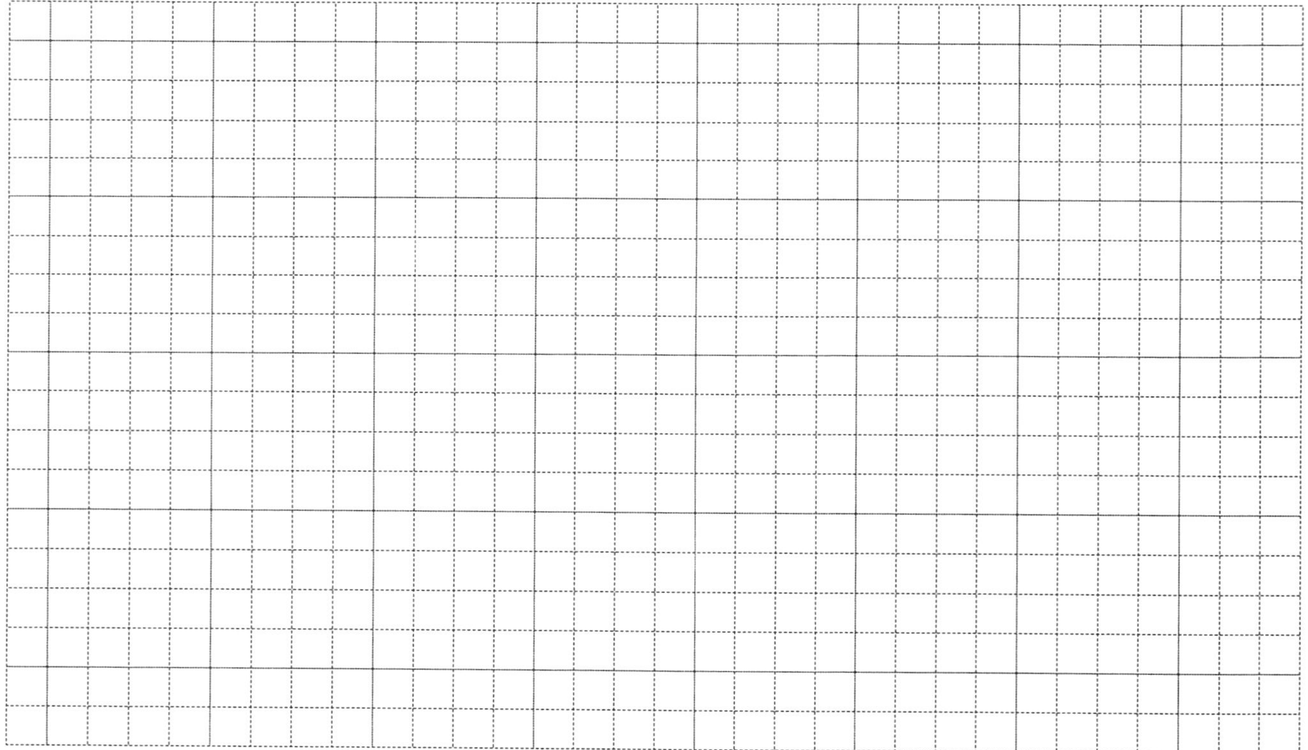

(*thAct60*, the activation time constant for h when the membrane potential is −60 mV; *thAct40*, the activation time constant when the membrane potential is −40 mV). You can verify that cycle period decreases as these time constants are decreased, demonstrating that the time constants for h activation, together with other circuit parameters (such as excitation as we have already seen), set the cycle period of the RI oscillator.

Circuit-4 Reciprocal inhibition oscillator: delayed excitation

For this exercise CIRCUIT is configured with three inhibitory neurons that are synaptically interconnected to simulate the oscillatory circuit that generates the swimming rhythm of the marine mollusc *Tritonia*. In this simulation N1 simulates the DSIs, N2 simulates the VSIs, and N3 simulates the C2s (fig. I.7-8a). Thus, N1 and N2 are interconnected by reciprocal, non-fatiguing, inhibitory synapses; N1 makes an excitatory synapse with N3; N3 provides excitatory synaptic drive to N2; and N2, in addition to its inhibitory synapse with N1, also makes an excitatory synapse with N3. The inhibitory synapses are fast-acting (decay time constant is about 50 ms), whereas the excitatory synapses all are very slow (decay time constant about 1 s). All three neurons also have an A conductance, which slows the rate at which they depolarize when an excitatory input is applied. Three TIMESERIES windows are open for this exercise, one for the membrane potential of each neuron. These windows are arranged to suggest the interaction loop with *Vm1* (potential of DSI) graphed at the upper left, *Vm2* (potential of VSI) graphed at the upper right, and *Vm3* (potential of C2) graphed at the lower right in figure I.7-8b. Two additional windows are open, the PARAMETER window to control model parameters and the STIMULATOR window to control the excitatory drive to the DSIs (N1).

The aim of this exercise is to illustrate how the *Tritonia* neuronal circuit generates the oscillations that underlie swimming movements in this animal. The fundamental mechanism for generating the oscillations in this circuit again is inhibition between two sets of neurons, the DSIs and VSIs (simulated by N1 and N2). The dynamics of the oscillations are inherent in the time delay by which DSI excitation is passed via C2 (N3) to activate the VSI (N2) neurons.

When this exercise begins (set the skip factor to 5 and then click on the STOP/GO icon), the three neurons in this circuit are quiescent, below impulse threshold. Swimming activity in *Tritonia* is normally initiated by massive sensory input to the oscillator DSI neurons. In this simulation, you can provide excitatory drive to N1 by injecting positive current (set *Stim1* to 1 and turn on the stimulator). Observe that the N1 impulse activity driven by the stimulator current hyperpolarizes the N2 neuron, in addition the excitatory synaptic connection from N1 to N3 (DSI to C2) slowly depolarizes N3 via temporal summation. With sufficient summation, the excitation from N1 raises the N3 membrane potential above threshold. Once active, N3 drives N2, which slowly depolarizes. The temporal summation of N3 input eventually depolarizes N2 above threshold, despite the continuing inhibition from N1. Now N2 inhibits both N1 and N3, which almost immediately cease generating impulses. Although inhibition of N3 terminates the excitatory input from N3 to N2, the slow nature of the excitatory input from N3 keeps N2 depolarized for about 1 s. When this excitation has finally waned, N2 ceases to fire, releasing N1 from inhibition. Soon N1 resumes its impulse activity, initiating a new cycle. Measure the period of the oscillations.

The normal swim episode in *Tritonia* is initiated, as we have just seen, with very strong excitatory drive to the DSI neurons. In the animal, this excitation gradually wanes, leading gradually to an increasing swim period until swimming stops after five to ten cycles. To study such a swim episode in this simulation first turn the stimulator off, use the CLOCK reset button to "zero" the trace. Contract the abscissa of each graph window for a full-scale time display of 40 s. Turn the stimulator on (set to 4 nA), allow the graph to plot two cycles, then step the current down to 3 nA. Again plot two cycles and set the current down to 2 nA. Repeat once more. Oscillations will now cease because of the lack of sufficient excitation to bring N2 (cell C2) above threshold. Note the increased cycle period when the excitation to the DSIs is decreased.

Section II Descriptions of the Models

This section provides an overview of each of the six models beginning with a brief introduction, glossaries of variable names and parameter names, and figures to illustrate the conceptual models and their implementation with *NeuroDynamix*. The description of the equations employed in the simulation and the relationships between variables and parameters can be found in section IV. This set of models is designed to simulate the dynamic properties of neurons at four levels of neuronal organization: the membrane patch, individual neuronal compartments (soma and axon), multicompartmental neurons, and neuronal circuits. The first model, RESISTOR, provides an introduction to electrical terms, concepts, and equations.

Section II.1 RESISTOR Model

RESISTOR simulates the properties of simple electrical circuits that include resistors, batteries, and capacitors. The experimenter can specify the number and value of components and then examine the consequence of applying known currents or known potentials to these circuits (fig. II.1-1). The list of variables and the default values of model parameters for RESISTOR are displayed in figure II.1-2.

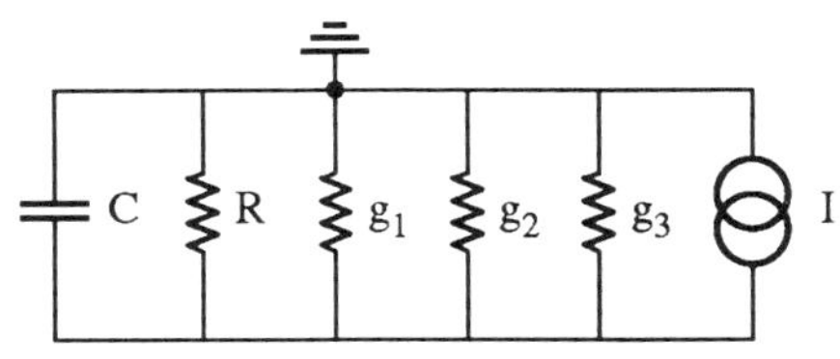

Figure II.1-1. Simple electrical circuit. This parallel conductance circuit includes a capacitor and four resistors, labelled *R*, g_1, g_2, and g_3. Current is passed through the circuit by the constant current source.

II.1.1 GLOSSARY OF GRAPHED VARIABLES (UNITS)

Vtot (mV) voltage as measured across the circuit

Itot (nA) total current passing through the circuit

Iresist (nA) current through the resistor

Ig1 (nA) current through conductance *g1*

Ig2 (nA) current through conductance *g2*

Ig3 (nA) current through conductance *g3*

Icap (nA) current through the capacitor (displacement current)

Stimulus (mV/nA) stimulus applied to the circuit. This may be a current or a potential, depending on the setting of the parameter *I/V* (see below).

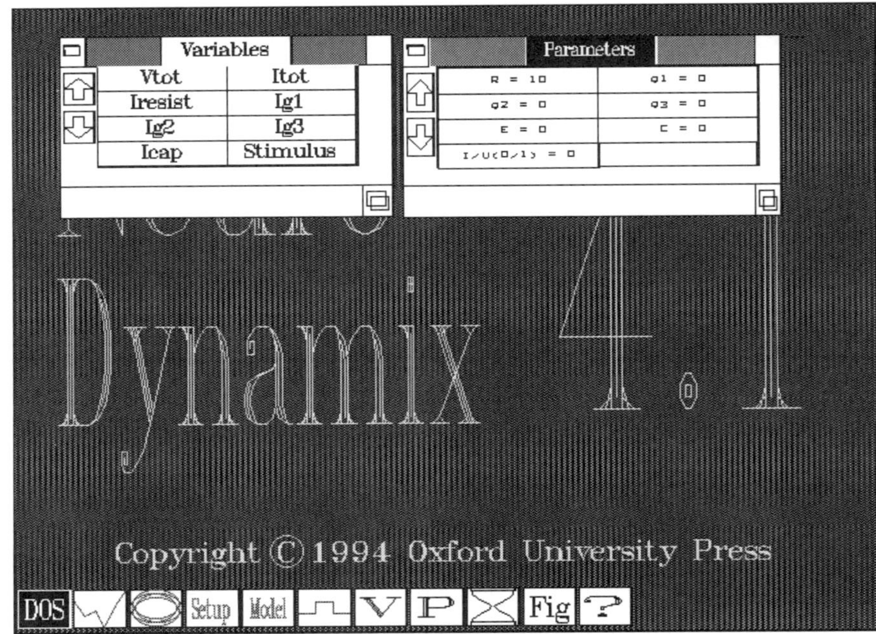

Figure II.1-2. VARIABLE and PARAMETER windows for the RESISTOR model.

II.1.2. GLOSSARY OF PARAMETERS (UNITS)

R (MΩ) — resistance of the resistor

$g1$ (μS) — conductance value for conductance 1

g2 (μS) — conductance value for conductance 2

$g3$ (μS) — conductance value for conductance 3

E (mV) — value of the battery potential

C (μF) — value of the capacitor

I/V (0/1) — determines whether stimulator applies a current ($I/V = 0$) or voltage ($I/V = 1$) to the circuit

Section II.2 PATCH Model

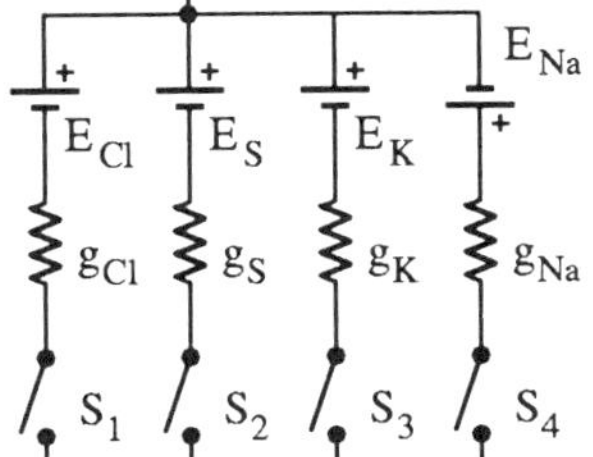

Figure II.2-1. Equivalent circuit for a membrane patch with four channels. These channels are from left to right, a chloride channel, an ACh channel, a potassium channel, and a sodium channel. Each of the resistors that symbolize these channels are shown in series with a switch that illustrates whether the channel is open or closed.

PATCH simulates the dynamics of a small piece of cell membrane as observed with the patch-clamp technique. Either individual channels in a small patch of membrane or the macroscopic current (up to 100 individual channels) observed with the whole cell patch clamp may be simulated. Channels are viewed as all-or-none conductances that are gated by voltage (sodium and potassium channels), by a messenger ligand (ACh channel), or not gated (chloride channel). The experimenter can specify the channel type (or types) and the number of channels found in the membrane patch. This modeling program does not include any time dependence for the gating processes; also, the sodium channel modeled here does not inactivate.

Properties that are simulated include the random opening of channels, the gating of the channels by membrane potential, the voltage dependence of the currents through the channels, the dependence of single-channel currents on channel conductance, and the summation of microscopic currents from many individual, randomly gated channels to yield a macroscopic current (fig. II.2-1). The list of variables and the default values of model parameters for PATCH are displayed in figure II.2-2.

II.2.1 GLOSSARY OF GRAPHED VARIABLES (UNITS)

INa (pA)	sodium current through the membrane patch
IK (pA)	potassium current through the membrane patch
ICl (pA)	chloride current through the membrane patch
ISyn (pA)	synaptic current through the membrane patch
ITotal (pA)	total current through the membrane patch
Vm (mV)	membrane potential; set by user
CACh	concentration of ACh applied to the patch for activating synaptic channels; applied by applying a positive stimulator current

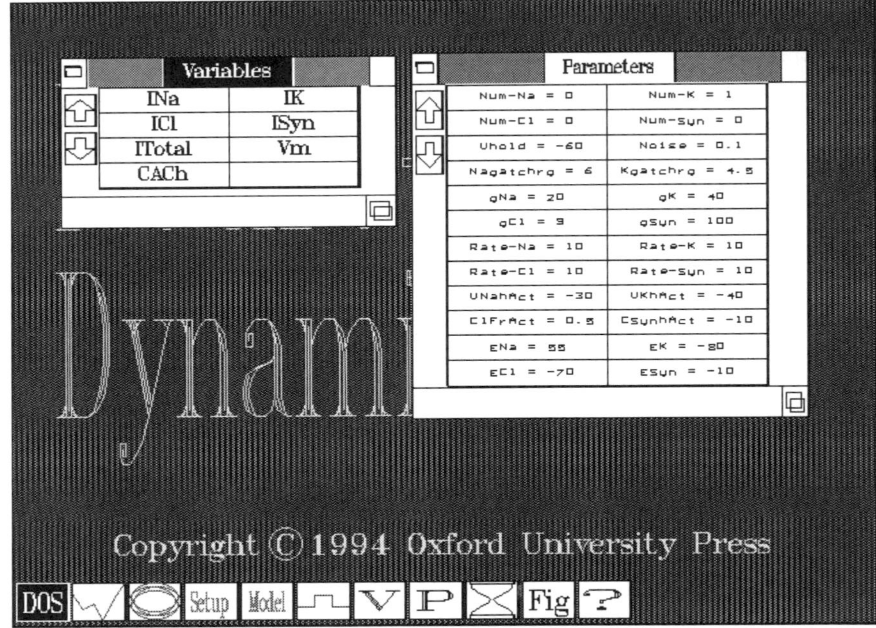

Figure II.2-2. VARIABLE and PARAMETER windows for the PATCH model.

II.2.2 GLOSSARY OF PARAMETERS (UNITS)

Num-Na (#)	number of sodium channels in membrane patch
Num-K (#)	number of potassium channels in membrane patch
Num-Cl (#)	number of chloride channels in membrane patch
Num-Syn (#)	number of synaptic channels in membrane patch
Vhold (mV)	holding potential across the membrane patch
Noise (mV)	amplitude of random noise added to the membrane potential
Nagatchrg (#)	number of assumed gating charges for sodium channels
Kgatchrg (#)	number of assumed gating charges for potassium channels
gNa (pS)	single-channel conductance of sodium channels
gK (pS)	single-channel conductance of potassium channels
gCl (pS)	single-channel conductance of chloride channels
gSyn (pS)	single-channel conductance of synaptic channels
Rate-Na (s^{-1})	average number of transitions, per second, from closed to open for individual sodium channels
Rate-K (s^{-1})	average number of channel transitions, per second, from closed to open for individual potassium channels

Rate-Cl (s^{-1})	average number of transitions, per second, from closed to open for individual chloride channels
Rate-Syn (s^{-1})	average number of transitions, per second, from closed to open for individual synaptic channels
VNahAct (mV)	membrane potential at which sodium channels are open 50% of the time
VKhAct (mV)	membrane potential at which potassium channels are open 50% of the time
ClFrAct (decimal)	fraction of time chloride channels are open
CSynhAct	transmitter concentration at which synaptic channels are open 50% of the time
ENa (mV)	sodium equilibrium potential
EK (mV)	potassium equilibrium potential
ECl (mV)	chloride equilibrium potential
ESyn (mV)	synaptic reversal potential

Section II.3 SOMA Model

SOMA is meant to simulate the origin of the resting potential in nerve cells. Because this model demonstrates steady-state conditions, or slow changes in the steady state, the membrane capacitance is ignored. In addition, there are no voltage- or time-dependent membrane conductances in this model. The equilibrium (Nernst) potentials for sodium, potassium, calcium, and chloride ions are calculated from the relevant ionic concentrations set in the PARAMETER window. The relationship between temperature and equilibrium potentials can be explored by altering the simulated temperature. When more than one ion has a nonzero conductance, the model calculates the resting potential via the parallel conductance model.

This model also includes a sodium/potassium pump, which generates a net outflow of positive ions and therefore acts to hyperpolarize the membrane. The assumed volume of the soma is 10 picoliter. The amount of sodium in the soma is controlled by the sodium influx and the pump rate. The amplitude of the pump current is determined by the internal sodium concentration, the maximum pumping rate, and the sensitivity of the pump to the intracellular sodium concentration. Please note that although the pump formally transports potassium as well as sodium ions, its effects on potassium concentrations are ignored in this simulation. Sodium ions may be injected into the cell to explore how sodium loading activates the pump. The pump is activated whenever the maximum pump rate is nonzero (fig. II.3-1). The list of variables and the default values of model parameters for SOMA are displayed in figure II.3-2.

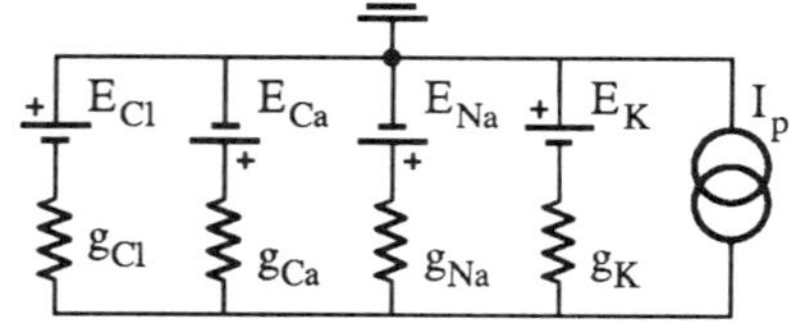

Figure II.3-1. Equivalent circuit for the SOMA model. Electrical symbols designate the Nernst potentials associated with each of the four types of channel conductances.

II.3.1 GLOSSARY OF GRAPHED VARIABLES (UNITS)

Vm (mV)	membrane potential
LCNaOut	log (base 10) of the extracellular sodium ion concentration
LCNaIn	log (base 10) of the intracellular sodium ion concentration

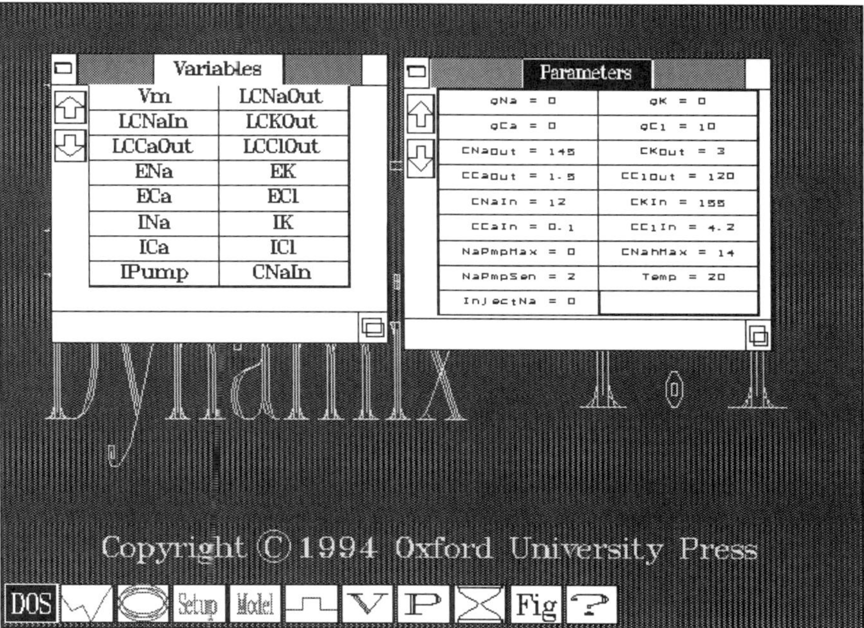

Figure II.3-2. VARIABLE and PARAMETER windows for the SOMA model.

LCKOut log (base 10) of the extracellular potassium ion concentration
LCCaOut log (base 10) of the extracellular calcium ion concentration
LCClOut log (base 10) of the extracellular chloride ion concentration
ENa (mV) equilibrium potential for sodium ions
EK (mV) equilibrium potential for potassium ions
ECa (mV) equilibrium potential for calcium ions
ECl (mV) equilibrium potential for chloride ions
INa (nA) sodium current
IK (nA) potassium current
ICa (nA) calcium current
ICl (nA) chloride current
IPump (nA) current generated by the sodium/potassium pump; the sign is positive for the net transport of positive ions out of the cell
CNaIn (mM) intracellular sodium ion concentration
Itot (nA) total membrane current
IStim (nA) current supplied by the stimulator

II.3.2 GLOSSARY OF PARAMETERS (UNITS)

gNa (nS)	sodium conductance
gK (nS)	potassium conductance
gCa (nS)	calcium conductance
gCl (nS)	chloride conductance
CNaOut (mM)	external sodium concentration
CKOut (mM)	external potassium concentration
CCaOut (mM)	external calcium concentration
CClOut (mM)	external chloride concentration
CNaIn (mM)	internal sodium concentration
CKIn (mM)	internal potassium concentration
CCaIn (μM)	internal calcium concentration
CClIn (mM)	internal chloride concentration
NaPmpMax (fmole/s)	maximum rate for the sodium pump
CNahMax (mM)	intracellular sodium concentration at which the sodium pump rate is at 0.5 of its maximum value
NaPmpSen (mM)	sensitivity of the pump to intracellular sodium concentration. Sensitivity increases as this parameter decreases.
Temp (°C)	temperature of the simulated soma
InjectNa (0/1)	set stimulator to inject sodium ions (*InjectNa* = 1; units are fmole/s) into the soma

Section II.4 AXON Model

AXON simulates the equations and parameters derived from experiments on the squid giant axon by Hodgkin and Huxley. Because of the exact correspondence between the equations incorporated into this model and the equations developed in the modeling studies of Hodgkin and Huxley, this model generates graphs that mirror precisely the theoretical curves depicted in the Hodgkin-Huxley papers on the squid giant axon. The space-clamped axon impulses calculated by Hodgkin and Huxley also are recreated in this model. The aim of this model, then, is to replicate Hodgkin and Huxley's parallel conductance model in order to illustrate their experimental data and to examine the implications of their theoretical conclusions (fig. II.4-1). The list of variables and the default values of model parameters for AXON are displayed in figure II.4-2.

The model can perform simulations in either of two experimental modes. In voltage-clamp mode, the model provides a means for recording total membrane currents generated by controlled, step changes in the membrane potential. In the current-clamp mode, the system provides a means of recording the membrane potential changes induced by *current* steps.

Specific properties illustrated with this simulation include 1) the time and voltage dependence of sodium and potassium currents and conductances; 2) the effect of temperature on rate constants; 3) the relationship between conductance changes and shape of the axon impulse; and 4) the mechanisms that underlie the generation of nerve impulses, repetitive firing, threshold, refractory period, and accommodation.

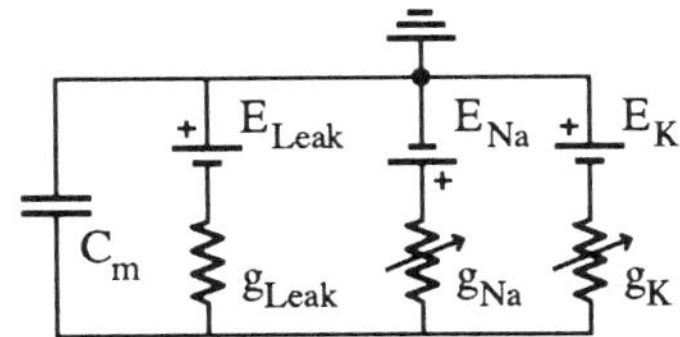

Figure II.4-1. Equivalent circuit for the AXON model. There are three ionic conductance paths through the membrane: a leakage constant conductance, a variable sodium conductance, and a potassium conductance.

II.4.1 GLOSSARY OF GRAPHED VARIABLES (UNITS)

INa (mA/cm^2 × 10) sodium current through the axon membrane

IK (mA/cm^2 × 10) potassium current through the axon membrane

ILeak (mA/cm^2 × 10) leakage current through the axon membrane

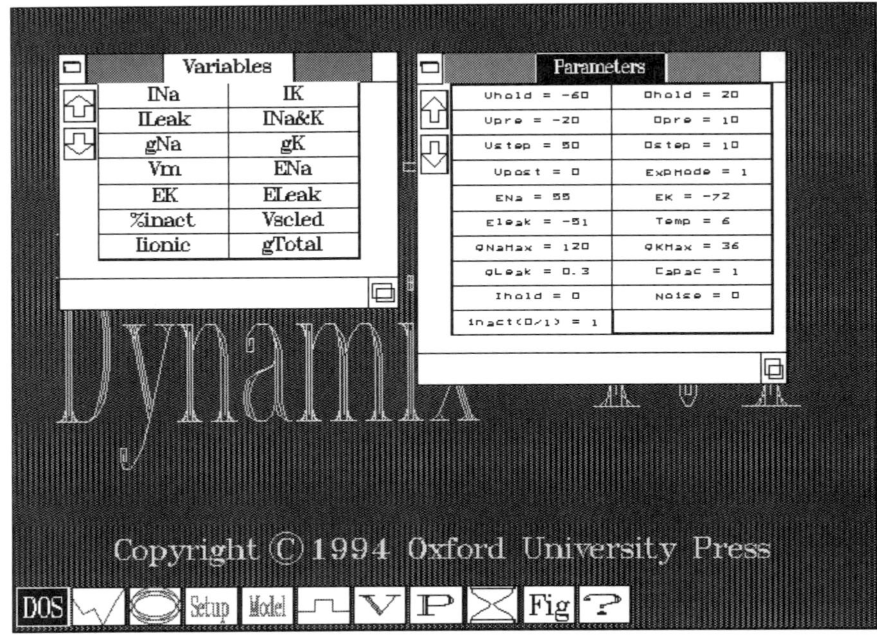

Figure II.4-2. VARIABLE and PARAMETER windows for the AXON model.

INa&K (mA/cm^2 × 10) sum of the sodium and potassium currents

gNa (mS/cm^2) calculated conductance of the axon membrane for sodium ions

gK (mS/cm^2) calculated conductance of the axon membrane for potassium ions

Vm (mV) membrane potential; set by the user in voltage-clamp mode, calculated in current-clamp mode

ENa (mV) equilibrium potential for sodium current, set on the parameter grid

EK (mV) equilibrium potential for potassium current, set on the parameter grid

ELeak (mV) equilibrium potential for the leakage current, set on the parameter grid

%inact percent inactivation of sodium conductance: $(1 - h) \times 100$

Vscled membrane potential offset and scaled by a factor of 0.1; for plotting membrane potential and current on the same graph

Iionic ($mA/cm^2 \times 10$) — total ionic current through the axon; the sum of *INa*, *IK,* and *ILeak*

gTotal (mS/cm^2) — sum of sodium, potassium, and leakage conductances

II.4.2 GLOSSARY OF PARAMETERS (UNITS)

Vhold (mV) — the membrane potential (holding potential) of the axon from time 0 for a duration determined by *Dhold* (parameter functions only in voltage-clamp mode)

Dhold (ms) — the duration of the interval during which the membrane potential is held at *Vhold* (parameter functions only in voltage-clamp mode)

Vpre (mV) — the membrane potential step (change from holding potential) of the axon from time *Dhold* for a duration determined by *Dpre* (parameter functions only in voltage-clamp mode)

Dpre (ms) — the duration of the interval (prepulse) during which the membrane potential is controlled by *Vpre* (parameter functions only in voltage-clamp mode)

Vstep (mV) — the membrane potential step (change from holding potential) following *Vpre* for a duration determined by *Dstep* (parameter functions only in voltage-clamp mode)

Dstep (ms) — the duration of the voltage step *Vstep* (parameter functions only in voltage-clamp mode)

Vpost (mV) — the membrane potential step (change from holding potential) following *Vstep*, this step is maintained until the trace reaches the right end of the graph (functions only in voltage-clamp mode)

Expmode (0/1) — selects the experimental mode, choices are current clamp (0) and voltage clamp (1)

ENa (mV) — sodium equilibrium potential, value is graphed as a variable

EK (mV) — potassium equilibrium potential, value is graphed as a variable

ELeak (mV) — leakage reversal potential, value is graphed as a variable

Temp (°C) — temperature at which the experiment is conducted

gNaMax (mS/cm^2) — maximum value of the sodium conductance per unit area. This is a measure of the density of sodium channels in the squid axon.

gKMax (mS/cm^2)	maximum value of the potassium conductance per unit area. This is a measure of the density of potassium channels in the squid axon.
gLeak (mS/cm^2)	value of the leakage conductance per unit area. This is a measure of the density of non–voltage-gated channels in the squid axon.
Capac ($\mu F/cm^2$)	specific capacitance of the axon membrane
Ihold (mA/cm^2)	value of constant, injected current
Noise (Mv)	amplitude of random noise added to V_m, the membrane potential
inact (0/1)	controls inactivation of sodium conductance, with inactivation either off (0) or on (1)

Section II.5 NEURON Model

NEURON simulates the membrane potentials in a neuron modeled as three serially connected compartments: dendrite, soma, and axon. The dendrite is represented here as a simple, single-compartment structure with only a leakage conductance and the usual membrane capacitance. The membrane potential associated with synaptic inputs from presynaptic neurons is simulated through by current injection by the experimenter, either into the dendrite or the soma compartments, from the stimulator.

The soma includes a variety of conductances and membrane pumps. This compartment is under direct experimental control (through the injection of transmembrane current). Three types of voltage-gated potassium currents are included: K (delayed rectifier) current, A current, and IR (inward or anomalous rectifier) current. Activation in all three currents is time dependent; inactivation rates for the K and A conductances also are time dependent. The soma compartment also includes a slow sodium current with time-dependent activation and inactivation. The simulated h current, activated on hyperpolarization, can be used to simulate both sag potentials on hyperpolarization and postinhibitory rebound. Finally the soma includes an electrogenic sodium/potassium ion pump that transports two potassium ions into the soma for every three sodium ions pumped out. The volume of soma in this simulation is taken as 10 picoliter. As in the SOMA model the effects of this pump on potassium concentrations are not simulated.

Nerve impulses are generated, in the axon compartment only, through brief sequential increases in sodium and potassium conductances that roughly approximate axonal biophysics; however, there is not a detailed simulation of impulses. Because this is a three-compartment model, there are longitudinal currents that pass between the three compartments. These are controlled by setting intercompartmental conductance values. This model is designed to simulate two types of neuronal complexities: multiple ionic currents and interactions within multicompartmental neurons (fig. II.5-1). The list of

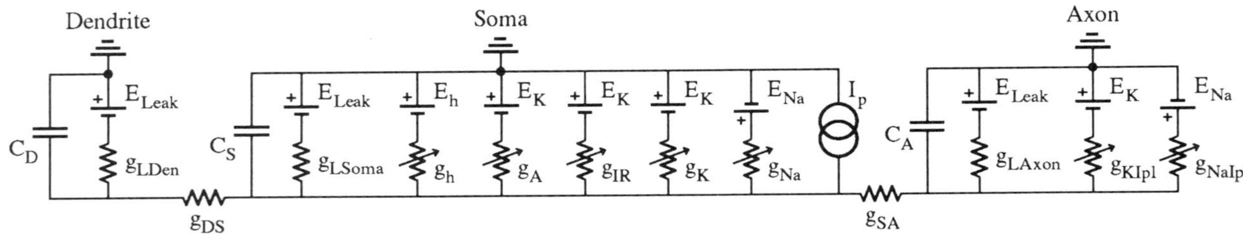

Figure II.5-1. Equivalent circuit for the three components of the NEURON model. The sodium conductance in the soma compartment simulates a slow sodium current. The sodium and potassium conductances shown for the axon component generate nerve impulses. The interlocking rings with the symbol I_P designate the soma sodium potassium pump.

variables and the default values of model parameters for NEURON are displayed in figure II.5-2.

II.5.1. GLOSSARY OF GRAPHED VARIABLES (UNITS)

VmD (mV)	dendrite membrane potential
VmS (mV)	soma membrane potential
VmA (mV)	axon membrane potential
gK (nS)	conductance for K (outward rectifier) current, in soma
gA (nS)	conductance for A current, in soma
gIR (nS)	conductance for inward rectifier current, in soma
gNa (nS)	conductance for inactivating sodium current, in soma
gh (nS)	conductance for h current (hyperpolarization-activated), in soma
IK (nA)	value of delayed rectifier current, soma
IA (nA)	value of A current, soma
IIR (nA)	value of inward rectifier current, soma
INa (nA)	value of sodium current, soma
Ih (nA)	value of h current, soma
INaPump (nA)	value of sodium pump current, soma
ILDen (nA)	value of leakage current in the dendrite
ILSoma (nA)	value of leakage current in the soma
CNaIn (mM)	intracellular concentration of sodium
ENa (mV)	sodium equilibrium potential

II.5.2 GLOSSARY OF PARAMETERS (UNITS)

Idc cell (nA)	amplitude of dc current injected into soma
StimD (1/0)	connects (1) or disconnects (0) stimulator output to the dendrite
StimS (1/0)	connects (1) or disconnects (0) stimulator output to the soma
gLDen (nS)	leakage conductance of the dendrite

(a)

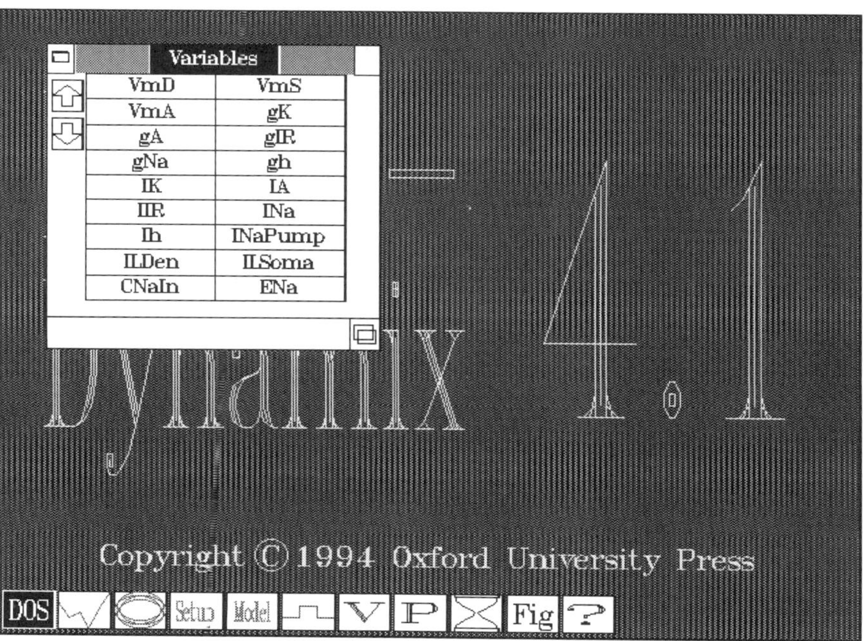

Figure II.5-2. VARIABLE (*a*) and PARAMETER (*b*) windows for the NEURON model.

(b)

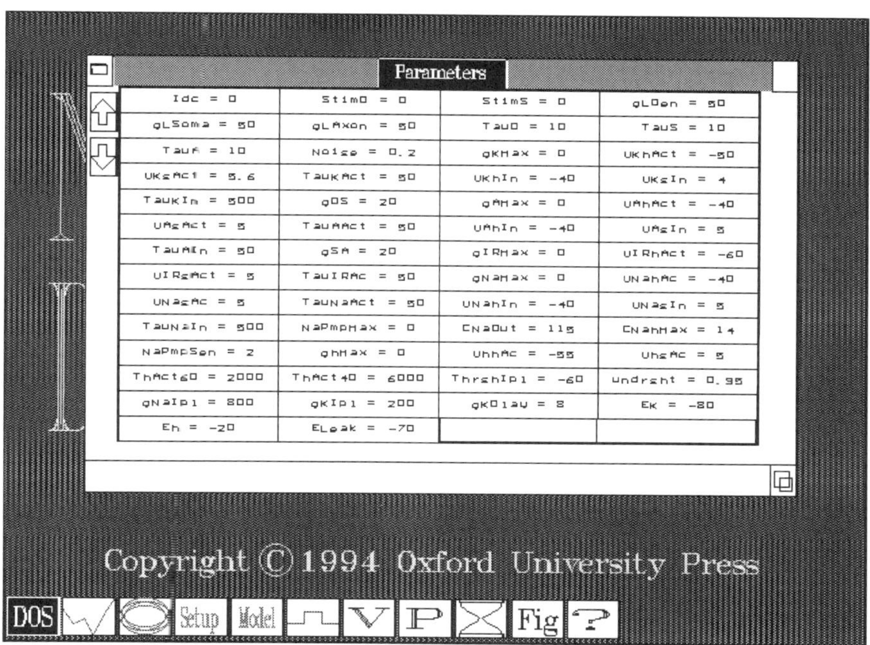

gLSoma (nS)	leakage conductance of the soma
gLAxon (nS)	leakage conductance of the axon
TauD (ms)	time constant of the dendrite
TauS (ms)	time constant of the soma
TauA (ms)	time constant of the axon
Noise (mV)	amplitude of noise added to *VmS*
gKMax (nS)	maximum value of the delayed rectifier conductance
VKhAct (mV)	membrane potential at which the delayed rectifier conductance is half activated
VKsAct (mV)	voltage sensitivity of the delayed rectifier conductance activation
TauKAct (ms)	time constant for activation of the delayed rectifier conductance
VKhIn (mV)	membrane potential at which the outward rectifier conductance is half inactivated
VKsIn (mV)	voltage sensitivity of outward rectifier conductance inactivation
TauKIn (ms)	time constant for inactivation of the delayed rectifier conductance
gDS (nS)	conductance of the coupling between dendrite and soma
gAMax (nS)	maximum value of A conductance
VAhAct (mV)	membrane potential at which the A conductance is half activated
VAsAct (mV)	voltage sensitivity of A conductance activation
TauAAct (ms)	time constant for A conductance activation
VAhIn (mV)	membrane potential at which the A conductance is half inactivated
VAsIn (mV)	voltage sensitivity of A conductance inactivation
TauAIn (ms)	time constant for A conductance inactivation
gSA (nS)	conductance of the coupling between soma and axon
gIRMax (nS)	maximum value of the inward rectifier conductance
VIRhAct (mV)	membrane potential at which the inward rectifier conductance is half activated
VIRsAct (mV)	voltage sensitivity of the inward rectifier conductance
TauIRAc (mV)	activation time constant for the inward rectifier conductance
gNaMax (nS)	maximum value of the sodium conductance. This is a slow, not impulse-related, sodium conductance in the soma.
VNahAc (mV)	membrane potential at which the slow sodium conductance is half activated

VNasAc (mV)	voltage sensitivity of the slow sodium conductance activation
TauNaAct (ms)	activation time constant for the sodium conductance
VNahIn (mV)	membrane potential at which the sodium conductance is half inactivated
VNasIn (mV)	voltage sensitivity of the slow sodium conductance inactivation
TauNaIn (ms)	time constant for the inactivation of the slow sodium conductance
NaPmpMax (fmoles/s)	maximum outward sodium ion flux generated by sodium pump. Maximum *net* flux is PumpMax/3.
CNaOut (mM)	extracellular concentration of sodium ions
CNahMax (mM)	intracellular concentration of sodium ions at which the sodium pump operates at 1/2 of the maximum rate
NaPmpSen (mM)	sensitivity of the sodium pump rate to changes in the concentration of intracellular sodium ions. When this parameter is small, the sensitivity is large.
ghMax (nS)	maximum value of the h conductance
VhhAc (mV)	membrane potential at which the h conductance is half activated
VhsAc (mV)	voltage sensitivity of the h conductance activation
ThAct60 (s)	time constant for h conductance activation when the membrane potential is −60 mV
ThAct40 (s)	time constant for h conductance activation when the membrane potential is −40 mV. Time constant at other potentials is via linear extrapolation.
ThrshIpl (mV)	membrane potential for the impulse threshold in the axon
undrsht (decimal)	multiplicative factor that decrements the potassium conductance during the undershoot of the nerve impulse. Increasing this number decreases impulse rate.
gNaIpl (nS)	maximum conductance for sodium during the impulse in the axon
gKIpl (nS)	maximum potassium conductance following the impulse in the axon
gKDlay (mS)	delay between onsets of the sodium and potassium conductances that give rise to the axon impulse
EK (mV)	potassium equilibrium potential
Eh (mV)	h conductance reversal potential
ELeak (mV)	reversal potential of the leakage conductance

Section II.6 CIRCUIT Model

CIRCUIT simulates the membrane potentials of a set of four one-compartment neurons. Each of these neurons includes elements from all three of the compartments described for NEURON, including simplified nerve impulses, a reduced set of membrane components, and experimenter-controlled currents. In addition, the CIRCUIT model includes synaptic conductances to simulate synaptic interactions between neurons and the properties of simple neuronal circuits. Simulated neuronal interactions can be of either the chemical synaptic variety (excitatory and inhibitory) or the electrical (rectifying and nonrectifying) variety.

In this model the types and magnitudes of synaptic interactions are under experimenter control. In addition, the inhibitory synapses can be set to exhibit synaptic fatigue based on a model in which transmitter depletion leads to reduced transmitter release from the presynaptic terminal and, hence, to reduced synaptic efficacy. The rate of transmitter depletion depends on the quantity of mobilized transmitter in the presynaptic terminal and on the membrane potential of the presynaptic axon. The electrotonic interaction is determined by interneuronal interactions in which the currents are set by the product of a conductance and the difference in the membrane potential between the interacting neurons. These interactions are simulated as rectifying, with pairs of such interactions simulating non-rectifying junctions. All possible chemical and electrotonic interactions between four neurons can be realized with this model. The time constants of both excitatory and inhibitory chemical interactions can be selected to simulate slow or fast synaptic interactions (fig. II.6-1). The list of variables and the default values of model parameters for CIRCUIT are displayed in figure II.6-2.

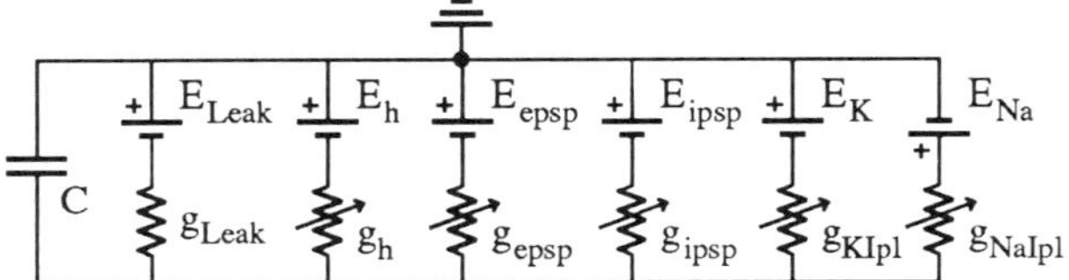

Figure II.6-1. Equivalent circuit for the CIRCUIT model. The sodium (G_{NaIpl}) and potassium (g_{KIpl}) generate impulses in this model.

(a)

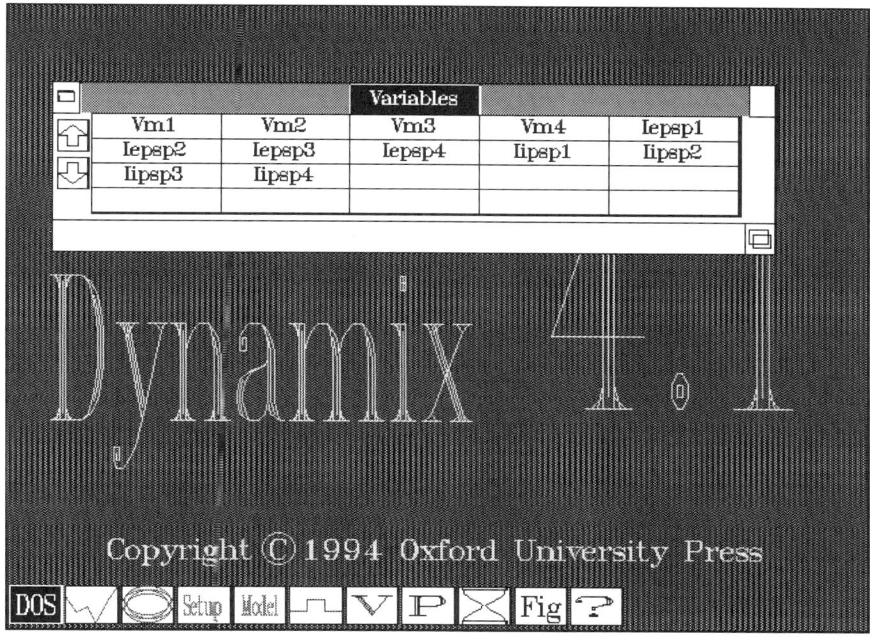

Figure II.6-2. VARIABLE (*a*) and PARAMETER (*b*) windows for the CIRCUIT model.

(b)

Parameters

Idc1 = 0	Idc2 = 0	Idc3 = 0	Idc4 = 0
Stim1 = 0	Stim2 = 0	Stim3 = 0	Stim4 = 0
Aex1>2 = 0	Aex2>1 = 0	Aex3>4 = 0	Aex4>3 = 0
Aex1>3 = 0	Aex3>1 = 0	Aex2>4 = 0	Aex4>2 = 0
Aex1>4 = 0	Aex4>1 = 0	Aex2>3 = 0	Aex3>2 = 0
Ain1>2 = 0	Ain2>1 = 0	Ain3>4 = 0	Ain4>3 = 0
Ain1>3 = 0	Ain3>1 = 0	Ain2>4 = 0	Ain4>2 = 0
Ain1>4 = 0	Ain4>1 = 0	Ain2>3 = 0	Ain3>2 = 0
gecp1>2 = 0	gecp2>1 = 0	gecp3>4 = 0	gecp4>3 = 0
gecp1>3 = 0	gecp3>1 = 0	gecp2>4 = 0	gecp4>2 = 0
gecp1>4 = 0	gecp4>1 = 0	gecp2>3 = 0	gecp3>2 = 0
ThrshSyn = -40	RateDepl = 100	RateRecv = 500	NumCells = 4
Eepsp = -10	TauEsP = 50	Eipsp = -70	TauIsp = 50
gAMax = 0	VAhAct = -40	VAsAct = 5	TauAAct = 50
VAhIn = -40	VAsIn = 5	TauAIn = 500	gLeak = 50
ghMax = 0	VhhAc = -55	VhsAc = 5	ThAct60 = 2000
ThAct40 = 6000	TauRest = 10	ThrshIpl = -40	undrsht = 0.9
gNaIpl = 800	gKIpl = 200	gKDlay = 4	Noise = 0.2
ENa = 55	EK = -80	Eh = -20	ELeak = -40

DOS Setup Model V P Fig

II.6.1 GLOSSARY OF GRAPHED VARIABLES (UNITS)

Vm1 (mV)	membrane potentials of simulated neuron N1
Vm2 (mV)	membrane potentials of simulated neuron N2
Vm3 (mV)	membrane potentials of simulated neuron N3
Vm4 (mV)	membrane potentials of simulated neuron N4
Iepsp1 (nA)	summed excitatory synaptic current for N1
Iepsp2 (nA)	summed excitatory synaptic current for N2
Iepsp3 (nA)	summed excitatory synaptic current for N3
Iepsp4 (nA)	summed excitatory synaptic current for N4
Iipsp1 (nA)	summed inhibitory synaptic current for N1
Iipsp2 (nA)	summed inhibitory synaptic current for N2
Iipsp3 (nA)	summed inhibitory synaptic current for N3
Iipsp4 (nA)	summed inhibitory synaptic current for N4

II.6.2 GLOSSARY OF PARAMETERS (UNITS)

Idc1 (nA)	amplitude of dc current injected into N1
Idc2 (nA)	amplitude of dc current injected into N2
Idc3 (nA)	amplitude of dc current injected into N3
Idc4 (nA)	amplitude of dc current injected into N4
Stim1 (1/0)	connects (1) or disconnects (0) stimulator output to N1
Stim2 (1/0)	connects (1) or disconnects (0) stimulator output to N2
Stim3 (1/0)	connects (1) or disconnects (0) stimulator output to N3
Stim4 (1/0)	connects (1) or disconnects (0) stimulator output to N4
Aexm > *n* (nS/mV)	the strength of the excitatory connection from neuron *m* to neuron *n*. There are 12 possible excitatory synaptic interactions.
Ainm > *n* (nS/mV)	the strength of the inhibitory connection from neuron *m* to neuron *n*. There are 12 possible inhibitory synaptic interactions.
gecpm > *n* (nS)	the conductance of the electrotonic connection (one direction only) from neuron *m* to neuron *n*. There are 12 possible rectifying interactions.
ThrshSyn (mV)	threshold membrane potential in presynaptic neurons. There is no synaptic effect if the membrane potential is more negative than this value.
RateDepl (ms^{-1} mV^{-1})	rate for presynaptic transmitter release
RateRecv (ms^{-1})	rate constant for replenishment of the transmitter in presynaptic neurons

NumCells (1–4)	number of cells in the simulation
Eepsp (mV)	reversal potential for excitatory synapses
TauEsp (ms)	time constant for the excitatory synaptic currents
Eipsp (mV)	reversal potential for inhibitory synapses
TauIsp (ms)	time constant of inhibitory synaptic currents
gAMax (nS)	maximum value of A conductance
VAhAct (mV)	membrane potential at which the A conductance is half activated
VAsAct (mV)	voltage sensitivity of A conductance activation
TauAAct (ms)	time constant for A conductance activation
VAhIn (mV)	membrane potential at which the A conductance is half inactivated
VAsIn (mV)	voltage sensitivity of A conductance inactivation
TauAIn (ms)	time constant for A conductance inactivation
gLeak (nS)	leakage conductance of all cells
ghMax (nS)	maximum value of the h conductance
VhhAc (mV)	membrane potential at which the h conductance is half activated
VhsAc (mV)	voltage sensitivity of the h conductance activation
thAct60 (s)	time constant for h conductance activation when the membrane potential is −60 mV
thAct40 (s)	time constant for h conductance activation when the membrane potential is −40 mV. Time constant at other potentials is via linear extrapolation.
TauRest (ms)	time constant for all neurons in the resting state
ThrshIpl (mV)	membrane potential for the impulse threshold in all neurons
undrsht (decimal)	multiplicative factor that decrements the potassium conductance during the undershoot of the nerve impulse. Increasing this number decreases impulse rate.
gNaIpl (nS)	maximum conductance for sodium during the impulse in the axon
gKIpl (nS)	maximum potassium conductance following the impulse in the axon
gKDlay (mS)	delay between onset of the sodium and potassium conductances that give rise to the axon impulse
Noise (mV)	amplitude of noise added to *Vm* in all neurons
ENa (mV)	sodium equilibrium potential
EK (mV)	potassium equilibrium potential

Eh (mV)	h conductance reversal potential
ELeak (mV)	reversal potential of the leakage conductance

Section III Numerical Methods

Section III.1 Form of the Equations

All of the neuronal models in *NeuroDynamix* are derived from the Hodgkin-Huxley-Katz parallel conductance model of the nerve cell membrane. The conductances through the membrane and between cell components give rise to ionic currents controlled by the values of the conductances and the electrochemical driving forces. Additional currents incorporated into some models are the displacement current through the membrane capacitor, those currents generated by ion pumps, and the currents applied from external sources. The fundamental assumption (Kirchhoff's rule) is that these currents must sum to zero at any node. All equations, for any compartment of the model, are of the form

$$I_e = I_b + I_m + I_s + I_p + I_C, \qquad \text{(III.1-1)}$$

where I_e represents all external currents (applied by the experimenter), I_b represents currents between the compartment under consideration and other compartments, I_m represents ionic currents through the membrane (except synaptic currents), I_s are all synaptic currents, I_p are pump currents, and I_C represents the capacitative current. Please note that each of these current types may include many specific currents. These currents are complex functions of time, membrane potential, and ionic concentration.

The current through the capacitor is given by

$$I_C = C\mathrm{d}V/\mathrm{d}t, \qquad \text{(III.1-2)}$$

where C is the value of the capacitor, V is the membrane potential and t is time. Hence, by rearranging equation (III.1-1) and substituting equation (III.1-2) we arrive at the fundamental equation underlying all of these simulations:

$$C\mathrm{d}V/\mathrm{d}t = -I_b - I_m - I_s + I_p + I_e. \qquad \text{(III.1-3)}$$

The ionic currents through channels may be written to express their dependence on the cell membrane potential explicitly. Therefore, we write (using Ohm's law)

$I_b = g_b(V - V_b)$ [current between present compartment and compartment b]
$I_m = g_m(V - E_m)$ [current through membrane conductance m]
$I_s = g_s(V - E_s)$ [current through synaptic conductance s],

where the g's are conductances, E_m is the Nernst equilibrium potential for a specific ionic channel, and E_s is the reversal potentials for a synaptic conductance. (Remember that for any specific system modeled there are many individual elements for each type of current.)

Then, substituting these explicit formulations into equation (III.1-3), we have

$$C\mathrm{d}V/\mathrm{d}t = -g_b(V - V_b) - g_m(V - E_m) - g_s(V - E_s) + I_p + I_e. \quad \text{(III.1-4)}$$

Collecting terms in V

$$C\mathrm{d}V/\mathrm{d}t = -(g_b + g_m + g_s)V + g_bV_b + g_mE_m + g_sE_s + I_p + I_e. \quad \text{(III.1-5)}$$

Let $G = g_b + g_m + g_s$, where G is the sum of all conductances in the neuron component under consideration; i.e., G is the total, instantaneous conductance. Then

$$\mathrm{d}V/\mathrm{d}t = (-GV + g_bV_b + g_mE_m + g_sE_s + I_p + I_e)/C. \quad \text{(III.1-6)}$$

Rearranging and dividing numerator and denominator by G yields

$$\frac{\mathrm{d}V}{\mathrm{d}t} = \frac{(g_bV_b + g_mE_m + g_sE_s + I_p + I_e)/G - V}{C/G}. \quad \text{(III.1-7)}$$

Finally, if we set

$$V_\infty = (g_bV_b + g_mE_m + g_sE_s + I_p + I_e)/G \text{ and } \tau = C/G, \quad \text{(III.1-8)}$$

where V_∞ is the steady potential (in the absence of changes in conductance values) and τ is the instantaneous membrane time constant, we have

$$\mathrm{d}V/\mathrm{d}t = (V_\infty - V)/\tau. \quad \text{(III.1-9)}$$

Equation (III.1-9) is the fundamental differential equation that must be solved numerically to determine the value of V at any time t, with the initial value of V (at $t = 0$) determined by the experimenter. With appropriate subscripts, V represents the membrane potential in any component of any cell. The specific form of V is determined by the ionic conductances, which themselves are functions of membrane potential (in the cell under consideration or in presynaptic neurons), and by external currents.

Section III.2 Numerical Solution

We seek a numerical solution to the equation

$$dV/dt = (V_\infty - V)/\tau \quad \text{(III.2-1)}$$

where V is a function of time, conductances, and external currents. The critical assumption for the solution proposed is that the dependence of V on variables other than time is weak or temporally slow. Using the substitutions

$$U = V_\infty - V \text{ and } dU = -dV \quad \text{(III.2-2)}$$

we have

$$dU/dt = -U/\tau. \quad \text{(III.2-3)}$$

The solution to this equation is

$$U = Be^{-t/\tau}, \quad \text{(III.2-4)}$$

where B is a constant. Then for small temporal increments, δ, we can write

$$U(t + \delta) = Be^{-(t+\delta)/\tau} = Be^{-t/\tau}e^{-\delta/\tau} = U(t)e^{-\delta/\tau}. \quad \text{(III.2-5)}$$

Changing variables back to V yields

$$V(t + \delta) = V_\infty + \{V - V_\infty\}e^{-\delta/\tau}. \quad \text{(III.2-6)}$$

Written in recursive form, equation (III.2-6) can be expressed as

$$V(j + 1) = V_\infty + \{V(j) - V_\infty\}e^{-\delta/\tau}, \quad \text{(III.2-7)}$$

where j is the iteration number and δ is the temporal increment between the jth and $(j + 1)^{th}$ iteration steps and where $V_\infty = (g_b V_b + g_m V_m + g_s E_s + I_p + I_e)/G$ and $\tau = C/G$. Equation (III.2-7) is an explicit description of the membrane potential as a function of time, provided that each iteration is a sufficiently small temporal increment of duration δ and provided that an initial value of $V(0)$ is known. All model equations that describe the changes of a variable as a function of time are of this form.

Section IV Equations Used in *NeuroDynamix* Simulations

This section provides a detailed description of the equations used in *NeuroDynamix* simulations as well as some justifications for the specific equations employed. The general form of the differential equations and their numerical integration are described in the section III, "Numerical Methods." Except for RESISTOR, all simulations are based on the parallel conductance model of Hodgkin, Huxley, and Katz. The limitations of this model at low ionic concentrations are also found in *NeuroDynamix*.

Section IV.1 Description of RESISTOR Equations

The most elementary equations in this model are based on Ohm's law, and, because there are no time-dependent variables, no numerical integration is required. Expressed in terms of resistance Ohm's law is simply

$$V = R * I, \tag{IV.1-1}$$

where V is the voltage, R is the resistance, and I is the current through the resistor. Its equivalent in conductance g (= $1/R$) is

$$I = g * V. \tag{IV.1-2}$$

For multiple conductors, this equation is extended as

$$V = I/(g_1 + g_2 + g_3). \tag{IV.1-3}$$

A final extension of Ohm's law is incorporated in a circuit consisting of a resistor and battery in series as

$$V = E + I/g, \tag{IV.1-4}$$

where E is the value of the battery potential.

Circuits consisting of a resistor in parallel with a capacitor connected to a constant-current source (fig. I.1-4) are described following the initiation of current injection at $t = 0$ by

$$V = V_0 * (1 - e^{-t/\tau}), \tag{IV.1-5}$$

where the time constant (τ) is equal to $R * C$ and whose final value, V_0, is $R * I$. When the current is turned off ($t = 0$), the equation describing the decline in the voltage across the circuit is

$$V = V_0 * e^{-t/\tau}. \qquad \text{(IV.1-6)}$$

Section IV.2 Description of PATCH Equations

For this simulation we picture the ionic channels as simple pores that have only two states, open and closed. The fundamental description for current flow of ions X through an individual channel is given by Ohm's law; namely,

$$I_X = g_X * (V_m - E_X), \tag{IV.2-1}$$

where I_X is the current through channel X (in picoamps); g_X is the channel conductance (in picosiemens); often designated by γ, V_m is the membrane potential; and E_X is the equilibrium potential for the ionic channel. The current in this equation is quantized because the conductance is assumed to have only two values, 0 (channel closed) or γ_X (channel open). The current equation can be rewritten using a Boolean variable "O"

$$I_X = \mathrm{O} * \gamma_X * (V_m - E_X), \tag{IV.2-2}$$

where $\mathrm{O} = 0$ when the channel is closed and $\mathrm{O} = 1$ when the channel is open.

IV.2.1 JUSTIFICATION FOR THE GATING EQUATIONS

The transition from the closed to the open state involves the movement of charge through the transmembrane electrical field associated with the membrane potential. Following the formulation of Hille, let w be the conformational energy increase associated with opening a channel in the absence of an electrical field. The electrical energy increase upon opening the channel in the presence of an electrical field is $-ZeV$, where Z is the number of charges moved, e is the charge on an electron, and V is the transmembrane potential. Then the change in energy consequent to channel opening is $w - ZeV$. The Boltzmann equation gives the ratio of open to closed channels at equilibrium as

$$\text{Open/Closed} = \exp[-(w - ZeV)/kT], \tag{IV.2-3}$$

where k is Boltzmann's constant and T is the absolute temperature. Rearrangement of this expression as the fraction of channels open (Open/Total) yields

$$\text{Open/Total} = \{1 + \exp[(w - ZeV)/kT]\}^{-1}. \tag{IV.2-4}$$

The value of Z (the effective charge translocated across the membrane during the conformational change associated with channel opening) is approximately 6 for the voltage-activated sodium channel and about 4.5 for the delayed rectifier potassium channel. At room temperature (20° C), e/kT is 25.26; hence, kT/eZ is 0.24 for sodium channels and 0.18 for delayed rectifier channels. Let f = the ratio Open/Total for some specific ion channel. For a large population of channels, half of the channels are open when $w = ZeV$. Therefore, we can write

$$f_{Na} = \{1 + \exp[0.24 * (V_{Nah} - V_m)]\}^{-1} \tag{IV.2-5}$$

and

$$f_K = \{1 + \exp[0.18 * (V_{Kh} - V_m)]\}^{-1}, \tag{IV.2-6}$$

where V_{Nah} is the potential at which half of the sodium channels in a population are open, V_{Kh} is the potential at which half of the potassium channels are open, and V_m is the membrane potential. Alternatively, we can view V_{Nah} and V_{Kh} as the potential at which sodium and potassium channels, respectively, are open 50 percent of the time.

The openings and closings of ionic channels are stochastic events with the fractions of open and closed channels determined by the membrane potential according to equations (IV.2-5 and IV.2-6). Because individual channels open and close independently, the ensemble equations given earlier can be replaced by temporal averages for single channels. Equations (IV.2-5 and IV.2-6) can be interpreted as the probability that a single channel will be open at any given time.

IV.2.2 PROBABILITY THAT A CHANNEL WILL OPEN OR CLOSE

In the implementation of the PATCH equations in *NeuroDynamix* time is quantized in steps that are 1/1000 of the full-scale time span of the abscissa. The probability that a given channel is open during any given integration step depends on both the probability that the channel will open (if closed) and on the probability that the channel will close (if open) during the step. Both of these probabilities are voltage dependent; i.e., channels are more likely to open (for most channels) and channels are less likely to close if the membrane potential is depolarized. The problem then is to derive equations that describe the probabilities of a channel opening and closing during any given step in terms of equations (IV.2-5) and (IV.2-6).

As above, let f be the fraction of the time that a given channel is open (this is equal to the fraction of channels open at any given time for an ensemble of identical channels). Now for a given number of time steps (N_T), f is equal to the number of channel openings (n_O) multiplied by the average open time (t_O) divided by the product of N_T and the duration (δ) of each step. That is

$$f = n_O * t_O/(N_T * \delta). \quad \text{(IV.2-7)}$$

But n_O is equal to the probability (P_O) that a closed channel will open during any given step multiplied by the number of steps (N_C) during which the channel is closed; hence,

$$f = P_O * N_C * t_O/(N_T * \delta). \quad \text{(IV.2-8)}$$

Because channels are either open or closed $N_C = N_T - N_O$; consequently,

$$f = P_O * (N_T - N_O) * t_O/(N_T * \delta). \quad \text{(IV.2-9)}$$

Dividing the numerator and denominator by N_T yields,

$$f = P_O * (1 - N_O/N_T) * t_O/\delta. \quad \text{(IV.2-10)}$$

But N_O/N_T is equal to f; therefore,

$$f = P_O * (1 - f) * t_O/\delta. \quad \text{(IV.2-11)}$$

Solving for P_O, we have

$$P_O = \delta * f/[(1 - f) * t_O]. \quad \text{(IV.2-12)}$$

Similarly, we can calculate P_C as

$$P_C = \delta * (1 - f)/(f * t_C). \quad \text{(IV.2-13)}$$

IV.2.3 PROBABILITY THAT A TRANSITION WILL OCCUR

In membrane patch recordings the number of transitions between channel open and closed states is determined by the membrane potential. The model employed here ignores this voltage dependence. Instead a parameter, whose value can be selected by the user, sets the average number of transitions per second from closed to open. Specific transition times are determined with a random number generator. Thus the number of transitions for any given time interval is not fixed. (But this number is independent of membrane potential in this simulation.)

Three types of ion channels are included in the PATCH model: voltage activated channels, a voltage-independent chloride channel, and a ligand-activated acetylcholine channel. The equations derived here strictly apply only to the voltage-activated channels. Because the chloride channel is not affected by membrane potential, the average open time is fixed, set by a model parameter. However, open and closed times are determined by a random number generator. The probability of opening for ACh channels follows the same considerations as those just described; however, the square of the acetylcholine concentration, instead of V_m, determines the average channel open time.

Section IV.3 Description of SOMA Equations

The SOMA model is meant to illustrate the origins of the resting potential; because current through the membrane capacitance is negligible at rest and during slow changes in membrane potential, we ignore the membrane capacitor in this simulation. In addition, none of the membrane conductances are voltage dependent.

The fundamental equation that describes the relationship between ionic concentrations and potential generated by a permeant ion X in a semipermeable membrane is that formulated by Nernst:

$$E_X = (RT/ZF) \ln \{[X]_1/[X]_2\}, \qquad \text{(IV.3-1)}$$

where R is the gas constant, T is the absolute temperature in Kelvin degrees, F is Faraday's constant, Z is the valence of the permeant ions, ln is the natural log, and X_1 and X_2 are the concentrations on the two sides of the membrane (outside and inside, respectively, for neurons).

When a cell includes channels that conduct several different ions (in this simulation sodium, potassium, calcium, and chloride), the resting membrane potential E_r is given by the weighted sum of the Nernst potentials, so that

$$E_r = E_{Na} * g_{Na}/g_T + E_K * g_K/g_T + E_{Ca} * g_{Ca}/g_T + E_{Cl} * g_{Cl}/g_T, \qquad \text{(IV.3-2)}$$

where g_X are the membrane conductances for individual ions and g_T is the total membrane conductance.

The SOMA model includes also an electrogenic sodium pump when the parameter *NaPmpMax* is set to a nonzero value. This pump simulates the extrusion of sodium ions, whose influx is due to the sodium conductance, thereby generating an outward transmembrane current. The amplitude of this current is set to 1/3 of the total ion transport rate because two potassium ions are translocated into neurons each time three sodium ions are expelled. (Note that the SOMA model ignores changes in the potassium concentration that might arise from pump activity.) The addition of the membrane pump adds an additional term to the equation describing the resting potential (IV.3-2); namely,

$$E_r = E_{Na} * g_{Na}/g_T + E_K * g_K/g_T + E_{Ca} * g_{Ca}/g_T + E_{Cl} * g_{Cl}/g_T - I_P/g_T + I_e/g_T, \qquad \text{(IV.3-3)}$$

where I_P is the net ionic current generated by the pump and I_e is current applied by the experimenter. The minus signs on the pump current term ensures that outward pump current hyperpolarizes the membrane.

Control of the pump rate for sodium ions is given by an equation similar to the Boltzmann equation for channel opening probabilities (as shown earlier):

$$J_P = J_{PMax}/\{1.0 + \exp[(C_h - C_i)/C_s]\}, \qquad \text{(IV.3-4)}$$

where J_P is the sodium ion flux generated by the pump, J_{PMax} is the maximum pump rate, C_h is the intracellular sodium concentration at which the pump rate is 0.5 of its maximum value, C_i is the intracellular sodium concentration, and C_s is the sensitivity of the pump to changes in intracellular sodium concentration (larger values generate lower sensitivities). Flux is converted to current via Faraday's constant F. Hence, the *net* ionic current generated by the pump is given by $I_P = 0.33 * F * J_P$.

Section IV.4 Description of AXON Equations

The AXON model is based on Hodgkin and Huxley's formulation of the parallel conductance model for the space-clamped squid axon. Most of the underpinnings for the AXON models are described in section I.4. Here we simply group the equations for the convenience of the interested reader.

The fundamental equation describing the flow of ions through squid axon membrane is

$$I_T = C\mathrm{d}V/\mathrm{d}t + I_i, \tag{IV.4-1}$$

where I_T is the total current density (current/unit area, outward current is taken as positive), C is the membrane capacitance per unit area, V is the membrane potential, t is time, and I_i is the total *ionic* current density. Following some rearrangement, with the individual ionic currents written explicitly and with the realization that under the experimental arrangements employed by Hodgkin and Huxley, the total membrane current is that applied by the experimenter (i.e., $I_T = I_e$), we have

$$C\mathrm{d}V/\mathrm{d}t = -I_{Na} - I_K - I_l + I_e, \tag{IV.4-2}$$

where I_{Na} is the sodium ion current density, I_K is the potassium ion current density, I_l is the leakage current density, and I_e is the current density applied externally by the experimenter.

Assuming that Ohm's law holds for the ionic currents, each of these currents is the product of a conductance and an electrochemical gradient:

$$I_{Na} = g_{Na}(V - E_{Na}), \tag{IV.4-3}$$

$$I_K = g_K(V - E_K), \text{ and} \tag{IV.4-4}$$

$$I_l = g_l(V - E_l), \quad \text{(IV.4-5)}$$

where E_{Na}, E_K, and g_l are the Nernst potentials for the sodium, potassium, and leakage conductances, respectively. The leakage conductance is constant, independent of membrane potential and time. The sodium and potassium conductances, however, are complex functions of time and membrane potential.

From their experimental results on the squid axon, Hodgkin and Huxley determined that the ionic conductances can be written as

$$g_{Na} = g_{Namax} m^3 h \text{ and} \quad \text{(IV.4-6)}$$

$$g_K = g_{Kmax} n^4, \quad \text{(IV.4-7)}$$

where g_{Namax} and g_{Kmax} are constants corresponding to the maximum values of membrane conductances for these ions and where m, h, and n are functions defined by the following first order differential equations:

$$\mathrm{d}m/\mathrm{d}t = \alpha_m(1 - m) - \beta_m = (m_\infty - m)/\tau_m, \quad \text{(IV.4-8)}$$

$$\mathrm{d}h/\mathrm{d}t = \alpha_h(1 - h) - \beta_h = (h_\infty - h)/\tau_h, \text{ and} \quad \text{(IV.4-9)}$$

$$\mathrm{d}n/\mathrm{d}t = \alpha_n(1 - n) - \beta_n = (n_\infty - n)/\tau_n. \quad \text{(IV.4-10)}$$

Here m and n are the activation factors for the sodium and potassium currents, respectively, whereas h represents the time course of sodium current inactivation. The solutions to these equations are

$$m = m_\infty - (m_\infty - m_0)\exp(-t/\tau_m), \quad \text{(IV.4-11)}$$

$$h = h_\infty - (h_\infty - h_0)\exp(-t/\tau_h), \text{ and} \quad \text{(IV.4-12)}$$

$$n = n_\infty - (n_\infty - n_0)\exp(-t/\tau_n), \quad \text{(IV.4-13)}$$

where

$$m_\infty = \alpha_m/(\alpha_m + \beta_m), \quad \text{(IV.4-14)}$$

$$h_\infty = \alpha_h/(\alpha_h + \beta_h), \quad \text{(IV.4-15)}$$

$$n_\infty = \alpha_n/(\alpha_n + \beta_n), \quad \text{(IV.4-16)}$$

and where

$$\tau_m = 1/(\alpha_m + \beta_m), \quad \text{(IV.4-17)}$$

$$\tau_h = 1/(\alpha_h + \beta_h), \text{ and} \quad \text{(IV.4-18)}$$

$$\tau_n = 1/(\alpha_n + \beta_n). \tag{IV.4-19}$$

The constants m_0, h_0, and n_0, are the values of m, h, and n, respectively, at $t = 0$, whereas m_∞, h_∞, and n_∞ are the steady-state (voltage-dependent) values of m, h, and n, achieved when t is very large.

Hodgkin and Huxley found explicit equations to describe m, h, and n by comparing their theoretical formulations with their voltage-clamp data. These equations are derived for a temperature of 6°C. The rate constants given here can be used for any temperature if they are temperature-corrected using a Q_{10} of 3.0 as follows:

$$\alpha_m = 0.1(25.0 + V_h - V)/[\exp(25.0 + V_h - V)/10.0 - 1.0], \tag{IV.4-20}$$

$$\beta_m = 4.0\exp(V_h - V)/18.0, \tag{IV.4-21}$$

$$\alpha_h = 0.07\exp(V_h - V)/20.0, \tag{IV.4-22}$$

$$\beta_h = 1.0/[\exp(30.0 + V_h - V)/10.0 + 1.0], \tag{IV.4-23}$$

$$\alpha_n = 0.01(10.0 + V_h - V)/[\exp(10.0 + V_h - V)/10.0 - 1.0], \tag{IV.4-24}$$

and

$$\beta_n = 0.125\exp(V_h - V)/80.0, \tag{IV.4-25}$$

where V_h is the holding membrane potential and V is the membrane potential at any time. In Hodgkin and Huxley's simulations as well as those in *NeuroDynamix*, m_0 and n_0 are set to 0 and h_0 is set to 1.0.

Section IV.5 Description of NEURON Equations

This NEURON model includes three compartments: dendrite, soma, and axon to simulate simple, spatially extended neurons. These compartments are autonomous except for intercompartmental currents that flow through electrotonic conduction pathways. The complete set of components included in the NEURON parallel conductance model is presented in figure I.5-1. The method for numerical integration employed throughout these simulations is satisfactory for the NEURON model except when the electrical conductance between compartments is high. In that special case, the equations become very stiff and the integration technique does not provide correct values; therefore, large values of the parameters g_{DS} and g_{SA} should be avoided.

For the dendrite component there are three parallel current paths through the membrane: 1) a resting conductance path, 2) a path for capacity current, and 3) current injected by the experimenter. The soma includes many additional ionic conductances and a membrane pump, whereas the axon includes only a leakage conductance and both fast sodium and potassium currents. Finally, there are two internal conductance paths, one that links the dendrite to the soma and a second that links the soma with the axon. The membrane potential of each component is determined by the ionic membrane conductances, the conductances between components, the membrane capacitance, the equilibrium potentials for membrane conductances, and the membrane potentials of adjacent components. All currents are expressed in nanoamperes (nA), potentials are in millivolts (mV), time is in seconds (s), and conductances are in nanosiemens (nS).

IV.5.1 DENDRITE EQUATIONS

The general form of the model equations for NEURON is that given in the section III, "Numerical Methods"; namely,

$$C\mathrm{d}V/\mathrm{d}t = -\Sigma I_X - \Sigma I_c + I_e, \tag{IV.5-1}$$

where $I_X = g_X(V - E_X)$ is the current through membrane conductance, X, $I_c = g_c(V_b - V_a)$ is the current between compartments b and a, and I_e is the external current. For a given compartment there may be several currents in each category.

For the dendrite compartment, which includes only a capacitor, a resting conductance and an external current source equation (IV.5-1) becomes

$$C_D\mathrm{d}V_D/\mathrm{d}t = -g_{LD}(V_D - E_r) + I_{eD} - I_{DS}, \tag{IV.5-2}$$

where $g_{LD}(V_D - E_L)$ is the current through the constant leakage (= resting) conductance g_{LD} of the dendrite, E_L is the resting potential, C_D is the capacitance, and I_{eD} is the external current applied by the experimenter. We can write explicitly that I_{DS}, the current between the neurite and the soma, is given by

$$I_{DS} = g_{DS}(V_D - V_S), \tag{IV.5-3}$$

where g_{DS} is reciprocal of resistance between the dendrite and the soma, V_D is the membrane potential of the dendrite, and V_S is the membrane potential of the soma. These equations, when combined, have the form required for numerical integration by the techniques outlined in section III.

IV.5.2 SOMA EQUATIONS

The equations for the soma compartment for the NEURON simulation are very similar to those of the dendrite compartment; however, in addition to the leakage conductance, the soma includes a delayed rectifier conductance, an A current, an inward rectifier conductance, a slow sodium current, the h current, and an electrogenic sodium pump. The additional conductances of the soma are both voltage and time dependent. The soma is electrotonically coupled to both of the other compartments.

For the soma compartment equation (IV.5-1) becomes

$$\begin{aligned} C_S\mathrm{d}V_S/\mathrm{d}t = {} & -g_{LS}(V_S - E_r) - \Sigma g_X(V_S - E_X) \\ & + I_{eS} + I_{DS} + I_{SA} - I_P, \end{aligned} \tag{IV.5-4}$$

where $g_{LS}(V_S - E_L)$ is the current through the leakage conductance g_{LS}, E_L is resting potential, C_S is the capacitance, and I_{eS} is external current applied by the experimenter. As for the dendrite, we have equations for I_{DS} and I_{SA}; namely,

$$I_{DS} = g_{DS}(V_D - V_S) \text{ and} \tag{IV.5-5}$$

$$I_{SA} = g_{SA}(V_A - V_S), \tag{IV.5-6}$$

where g_{SA} is reciprocal of resistance between soma and axon and V_A is the membrane potential of the axon. The equation for the pump current I_P is equivalent to the one developed earlier in section IV.3 for the SOMA model.

The term ΣI_X represents all of the time- and voltage-dependent currents of the soma compartment—I_K, I_A, I_{IR}, I_{Na}, and I_h. Each of these currents are generated by voltage-dependent conductance whose activation value Act_X is given by

$$Act_X = 1/\{1 + \exp[-(V - V_{XhAct})/V_{XsAct}]\}, \quad \text{(IV.5-7)}$$

where V is the membrane potential, V_{XhAct} is the membrane potential at which half of the channels are activated (channels are open half of the time), and V_{XsAct} is the voltage sensitivity of activation. The inverse of this last term is a measure of the number of charges moving through the membrane when a channel is gated open. (Note that for conductances that are activated by hyperpolarization the V_{XsAct} term is multiplied by -1).

For currents that also undergo inactivation (in this simulation: I_K, I_A, and I_{Na}), the corresponding inactivation term is

$$In_X = 1/\{1 + \exp[(V - V_{XhIn})/V_{XsIn}]\}, \quad \text{(IV.5-8)}$$

where V_{XhIn} is the membrane potential at which half of channels are inactivated and V_{XsIn} is the voltage sensitivity of inactivation.

All of the voltage-dependent channels in the NEURON model also are time dependent. Thus in each of the activation and inactivation processes described by equations (IV.5-7) and (IV.5-8) there also is an associated time constant. For I_K, I_A, I_{IR}, and I_{Na} these time constants are fixed, without any voltage dependence. The time constant for I_h activation is also voltage dependent, with the actual value of the time constant at any membrane potential determined by a linear interpolation between the time constants (set by model parameters) at $V_S = -40$ mV and -60 mV.

The time dependence of the activation Act_X for any conductance X is described by the differential equation

$$dAct_X/dt = (1 - Act_X)/\tau_{ActX}, \quad \text{(IV.5-9)}$$

where τ_{ActX} is the activation time constant. Similarly, the time dependence of inactivation is given by

$$dIn_X/dt = (1 - In_X)/\tau_{InX}, \quad \text{(IV.5-10)}$$

where the terms correspond to those of equation (IV.5-9). These equations are solved by numerical integration as described in section III.

Finally, the conductances that give rise to voltage- and time-dependent ionic currents in the NEURON model (and in the CIRCUIT model) are described by combining equations (IV.5-7), (IV.5-8), (IV.5-9), and (IV.5-10) as

$$g_X = Act_X * In_X * g_{Xmax}, \quad \text{(IV.5-11)}$$

where g_{Xmax} is the maximum value for conductance g_X.

IV.5.3 AXON EQUATIONS

The axon compartment includes a resting conductance path, a capacitative current, and an electrotonic conductance between this compartment and the neurite. In addition, there are fast, transient conductances for sodium and potassium ions to simulate nerve impulses without the computational overhead associated with the Hodgkin-Huxley equations. The basic form of the equation is similar to the other two compartments with

$$C_A dV_A/dt = -g_{LA}(V_A - E_L) - g_{Na}(V_A - E_{Na}) - g_K(V_A - e_K) - I_{SA}, \tag{IV.5-12}$$

where $g_{LA}(V_A - E_L)$ is the current through the leakage conductance g_{LD} of the axon, E_L is the resting potential, C_A is the capacitance, and where I_{SA}, the current between soma and axon, is given by

$$I_{SA} = g_{SA}(V_A - V_S), \tag{IV.5-13}$$

where g_{SA} is reciprocal of resistance between axon and soma.

The rising phase and overshoot of the nerve impulse is simulated by a step of g_{Na} to a large value (parameter *gNaIpl*). The depolarization engendered by this step change in sodium conductance is not instantaneous because of membrane capacitance. After a duration selected by the user (parameter *gKDlay*), the sodium conductance is reset to 0 and g_K is set to a large value (parameter *gKIpl*). This generates an undershoot whose amplitude is determined by *gKIpl*. Following its maximum value the undershoot decays because amplitude of g_K is decremented during each numerical integration step by multiplying its current value by a fraction (whose value is near 1.0; parameter *undrsht*).

Section IV.6 Description of CIRCUIT Equations

This simulation comprises four single-compartment neurons that interact via electrical connections (rectifying and nonrectifying) and chemical synapses (excitatory and inhibitory). The description below provides the equations for electrical coupling, inhibitory and excitatory synaptic interactions, and synaptic fatigue. The equations that underlie leakage current, A current, h current, and for simulating the nerve impulse and that are described in detail in section IV.5 are not repeated here. In both vertebrates and invertebrates synaptic transmission may be mediated by nerve impulses or by the tonic, voltage-modulated release of transmitter; both of these modes are included in the CIRCUIT model. In addition, synaptic interactions incorporated into CIRCUIT may include synaptic fatigue.

IV.6.1 EQUATIONS DESCRIBING INHIBITORY PRESYNAPTIC PROCESSES

Synaptic interactions in CIRCUIT are simulated as the product of two processes—the release of the transmitter from presynaptic terminals and the opening of synaptic channels by the transmitter in the postsynaptic membrane. We assume that the presynaptic terminals have the transmitter packaged in some number of vesicles available (mobilized) for immediate release. We also assume that this vesicle store is not only depleted during synaptic transmission but also continuously replenished until it reaches some maximum amount.

Let N_T be the total number of vesicles in the presynaptic terminal, f_T be the fraction of vesicles that are mobilized, k_d be the rate constant for transmitter depletion (release), and k_r be the rate for transmitter replenishment (reuptake, synthesis, or mobilization by Ca ions). We assume that the rate of transmitter depletion is $k_d * N_T * f_T * V_s$, where the last term is the difference between the membrane potential, V, and the threshold for release of transmitter V_0. (Note that V_s is set to 0 if V is less than V_0.) Then

$$N_T * df_T/dt = -k_d * V_s * N_T * f_T + N_T * k_r * (1 - f_T), \quad \text{(IV.6-1)}$$

where the first term on the right side of the equation is the rate of transmitter depletion and the second term is the rate at which the transmitter is replenished. This equation can be rearranged to yield

$$df_T/dt = (k_d * V_s + k_r) * [k_r/(k_d * V_s + k_r) - f_T]. \quad \text{(IV.6-2)}$$

Let $f_{T\infty} = k_r/(k_d * V_s + k_r)$ and $\tau = 1/(k_d * V_s + k_r)$. Then by substitution

$$df_T/dt = (f_{T\infty} - f_T)/\tau. \quad \text{(IV.6-3)}$$

This final equation is integrated numerically by the methods outlined in section III.

IV.6.2 EQUATIONS DESCRIBING INHIBITORY POSTSYNAPTIC PROCESSES

At directly acting synapses neurotransmitter released from the presynaptic membrane binds to ion channels in postsynaptic cells. For simplicity, we assume that there is a linear relationship between transmitter release and postsynaptic conductance increases. Hence, for inhibitory synapses the synaptic conductance g_i is given by

$$g_i = k_d * N_T * f_T * V_s, \quad \text{(IV.6-4)}$$

where the terms in the equation are as defined earlier. The first two terms on the right side of equation (IV.6-4) can be combined into a single parameter g_{is}, which is a measure of overall synaptic strength. Therefore, we can write

$$g_i = g_{is} * f_T * V_s, \quad \text{(IV.6-5)}$$

where f_T is defined by (IV.6-3). This variable, which ranges between 0 and 1, determines the synaptic fatigue of inhibitory synaptic interactions. The term V_s describes the depolarization of the presynaptic membrane beyond threshold for transmitter release.

The synaptic current I_{is} generated in the inhibited postsynaptic neuron by depolarization of the presynaptic terminal is given by

$$I_{is} = g_i * (V - E_i), \quad \text{(IV.6-6)}$$

where E_i is the reversal potential for the synaptic conductance and $(V - E_i)$ is the driving force for synaptic current.

IV.6.3 EXCITATORY SYNAPSES

Excitatory synaptic connections are modeled very simply, for they are non-fatiguing in this simulation. In particular the postsynaptic conductance g_e due to suprathreshold excitation of the presynaptic neuron is described by

$$g_e = g_{es} * V_s, \tag{IV.6-7}$$

where g_{es} is a measure of synaptic strength and V_s again is the depolarization of the presynaptic membrane beyond threshold for transmitter release.

The synaptic current I_{es} generated in the postsynaptic neuron by depolarization of the presynaptic terminal is given by

$$I_{es} = g_e * (V - E_e), \tag{IV.6-8}$$

where E_e is the reversal potential for the synaptic conductance and $(V - E_e)$ is the driving force for synaptic current.

IV.6.4 ELECTRICAL INTERACTIONS

Electrical interactions are simulated in CIRCUIT by directed, rectifying interactions. For example, if neurons n and m are coupled by a rectifying junction that allows current to pass from n to m, the interaction is described by the equation

$$I_{nm} = g_{nm} * (V_n - V_m), \tag{IV.6-9}$$

for V_n less negative than V_m, but $I_{nm} = 0$ otherwise. In this equation g_{nm} is the value of electrical conductance coupling the two neurons. This formulation allows current to pass in one direction only. Non-rectifying interactions are established between CIRCUIT neurons by implementing pairs of interactions described by equation (IV.6-9) with the terms n and m interchanged.

Incorporation of these electric junctional currents together with the synaptic, resting and impulse currents into the fundamental equation for the parallel conductance model yields the final equation that describes CIRCUIT neuron behavior:

$$\begin{aligned} C\mathrm{d}V/\mathrm{d}t = &-g_L * (V - E_L) - g_{Na} * (V - E_{Na}) - g_K * (V - E_K) \\ &- I_{es} - I_{is} - I_{nm} + I_e , \end{aligned} \tag{IV.6-10}$$

where g_L is the leakage conductance, V is the membrane potential, E_L is the resting potential, g_{Na} is the transient sodium conductance generating the nerve impulse, E_{Na} is the sodium equilibrium potential, g_K is the delayed rectifier current that generates the impulse undershoot, E_K is the potassium equilibrium potential, I_{es} is the excitatory synaptic current, I_{is} is the inhibitory synaptic current, I_{nm} is the current from cell n from cell m through an electrical junction, and I_e is the externally applied current. The numerical integration of this equation is described in section III.

Appendix: Guide to the *NeuroDynamix* Windows

A.1 INTRODUCTION AND INSTALLATION

The *NeuroDynamix* modeling program is written in Borland C++ 2.0 and runs on IBM compatible (80286, 80386, or 80486 based) computers equipped with at least 640K RAM, a math coprocessor, and an EGA or VGA graphics adaptor. A mouse is required to control all aspects of the program once it is loaded into memory. Although some functions of the program are executed much more quickly if the program files are loaded onto a fixed disk (better yet, a RAM disk), *NeuroDynamix* can be run from a floppy disk drive. (Please note: The programs will not execute properly if the disk is write-protected.) *NeuroDynamix* includes a set of six different models that simulate neuronal function from the level of individual ion channels, as observed with the patch clamp, to the level of a small neuronal circuit. Selection and execution of the individual models occurs from within a single program (ND.EXE). Additional files associated with *NeuroDynamix* include an installation file (INSTALL.BAT) for installing *NeuroDynamix* on a fixed disk; a setup file (NDSETUP.EXE) for changing some installation variables, such as the number of buttons on the mouse and colors on the monitor screen; a file that stores the configuration information generated by the setup program (ND.CNF); a set of six parameter files that provide specific default information for each of the models (extension .PRM); and one or more graphics driver files (extension .BGI). Please consult the README.TXT file for the latest update information.

The user has access to program execution via a graphical user interface designed specifically for this application. This interface consists of a set of structurally equivalent windows that allow the user to configure specific neuronal models (by choosing variables to be plotted, by setting parameter values, etc.) and then to observe the model outputs in a window that graphs model variables as a function of time or on the phase plane. Completely transparent to the user, the differential equations that comprise the models are solved numerically through small iterative temporal steps. This modeling system is intended to simulate closely the experimental procedures used to study the electrophysiology of nervous systems and to display the results as they are usually observed on an oscilloscope. Because the models are computationally intensive, the graphing rate is considerably slower than that encountered during electrophysiological experimentation on living nervous tissue. The extent of this temporal dilation depends critically on the speed of the computer employed in the simulation.

This section provides general information and instructions for 1) using the various windows; 2) controlling of model output interactively by manipulating displays and model parameters; and 3) initiating a modeling session. A "Quick Reference Guide" is included in this book. Specific, more-detailed descriptions for each of the models, including glossaries of variables and parameter names are found in section II.

A.1.1 Windows

A.1.1.1 Window types There are two types of windows: primary windows that are accessible via the icons at the bottom of the monitor screen and secondary windows that are displayed when activated from another window. The primary windows include the TIMESERIES, PHASEPLANE, SETUP, MODEL, STIMULATOR, VARIABLES, PARAMETERS, CLOCK, and FIGURE windows. These primary windows are opened (activated) by clicking the mouse once on the appropriate icon at the bottom of the screen (fig. A.1-1). These icons are active and are available at all times. The secondary windows are the PARAMETER MODIFICATION window (opened from the PARAMETERS and STIMULATOR windows) and the ANALYZE window (opened from the TIMESERIES AND PHASEPLANE windows).

Figure A.1-1. Monitor display of the initial *NeuroDynamix* window

A.1.1.2 Opening windows To open a primary window, click the mouse once on the appropriate icon. The window will usually appear in an unused area of the screen in its minimum, default size. The window can then be moved and resized as desired (described later). If no clear space is available, the window will appear overlaid on an existing window. Secondary windows are opened in special ways. To open a PARAMETER MODIFICATION window, click LEFT and RIGHT mouse buttons in rapid succession (or click the middle button on a 3-button mouse) on a specific parameter (in the PARAMETERS or STIMULATOR window). The PARAMETER MODIFICATION window permits rapid alterations in parameter and stimulator values. The ANALYZE window is opened by clicking the mouse once on the cross-hairs icon found at the lower margin of the TIMESERIES and PHASEPLANE windows.

A.1.1.3 Moving windows All windows can be moved anywhere on the monitor screen by depressing and holding the RIGHT mouse button on the title bar at the upper margin and dragging the window outline by moving the mouse. The window will relocate to the new position when the button is released. (If the window refuses to move as far as you wish, reset the mouse by moving the window to the opposite side of the screen, as far as it will go, and then slowly moving the window to the desired destination.)

A.1.1.4 Resizing windows The TIMESERIES, PHASEPLANE, MODEL, SETUP, VARIABLES, and PARAMETERS windows are resizable. They can be expanded from the minimum size to fill much of the screen. To resize a window, depress the mouse on the RESIZE icon (lower right-hand corner) and drag that corner to a new position (the upper left-hand corner remains stationary). Upon releasing the mouse, the window will be redrawn with the new size. You cannot reduce a window to less than its minimum (default) size.

A.1.1.5 Closing windows All windows can be closed (removed from the monitor screen) by clicking the mouse on the CLOSE icon (upper left-hand corner). When a window is reopened, its appearance is not related to the state of the window that appeared previously; that is, neither the size nor the position of a window is preserved once it is closed.

A.1.1.6 Scrolling Many of the windows contain lists of information (parameters, variables, models available, and setup specifications). If one of these windows is too small to view the entire list at once, information above or below the current window area can be viewed by scrolling. Clicking the LEFT mouse button on the up arrow (upper edge, left border) scrolls the window up one position; clicking the LEFT mouse button on the down arrow scrolls the window down one position. Scrolling can be accomplished in larger, window-length steps by clicking the RIGHT mouse button on the up or down arrow.

A.1.1.7 Minimizing The STIMULATOR window can be displayed in two forms. The large default form includes all available control parameters. A reduced or minimized form with limited information (and control capabilities) can be obtained by clicking the mouse on the MINIMIZE icon in the upper right-hand corner of the large STIMULATOR window. The full-sized window display for the stimulator can be regained by clicking on the MAXIMIZE icon in the upper right-hand corner of the reduced STIMULATOR window.

A.1.1.8 Window overlap When two windows overlap, the overlying window is functional and the lower, covered window is dormant. Thus updating of the covered window (even if only partially covered) does not occur. If, for example, a variable is dropped into a partially covered TIMESERIES window, nothing will seem to happen. However, the variable is now ready for graphing, as can be seen when the overlying window is removed (either by resizing it, by moving it, or by clicking the mouse anywhere on the lower window). To restore a partially overlaid, dormant window to the functional condition, click the mouse anywhere on the exposed window. The two windows will be redrawn in the same position but with the newly activated window on top. Note that the calculations to update the TIMESERIES and PHASEPLANE windows continue even when these windows are not active. It is advisable to overlap only those windows that are accessed infrequently. Note that the icons at the bottom of the screen remain active even if they are overlapped by one of the windows.

A.2 RUNNING *NEURODYNAMIX*

The following files must be located in one disk directory: ND.EXE, ND.CNF, EGAVGA.BGI (or some other graphics driver). The parameter files RESISTOR.PRM, PATCH.PRM, SOMA.PRM, AXON.PRM, NEURON.PRM, and CIRCUIT.PRM are necessary for performing the preconfigured modeling exercises provided with this system. If you need (or wish) to reconfigure the setup instruction to *NeuroDynamix,* the file NDSETUP.EXE also must be in this disk directory.

Before starting the program be sure that the mouse driver has been installed. If this is the first use of *NeuroDynamix* on a particular computer, it *may* be necessary to run the program NDSETUP.EXE to properly configure the modeling system. The setup program is self-contained and is supplied with helpful instructions. Its primary function is to ensure that *NeuroDynamix* is informed about the number of buttons on your mouse. Computer simulation is initiated by selecting the appropriate disk drive from the DOS command line and then entering *ND*. This initiates program execution and causes display of the initial monitor screen (fig. A.1-1). The series of icons at the lower left margin of the screen serves to direct program execution. Click the mouse on the MODEL icon to open the MODEL selection window. Then highlight and select the desired model by clicking the mouse once on the model name (scroll through the list to view all model names). Click the mouse a second time on the model name to open the SETUP selection window. Click the LEFT mouse button twice on the desired setup configuration to begin a specific modeling exercise. Finally, begin model execution (graphing variables) by clicking the mouse on the STOP/GO icon.

A.3 SPECIFIC DESCRIPTIONS OF *NEURODYNAMIX* WINDOWS

A.3.1 MODEL selection window

The first step in beginning a modeling session is to select a specific neuronal model. Models are selected by opening the MODEL window (click the mouse on the MODEL icon) at the beginning of a modeling session (fig. A.3-1). The default MODEL window displays four of the available models. The additional models are viewed either by scrolling or by enlarging the window. Click the mouse once on a model name to select a specific model, which will then become highlighted; click the mouse a second time to open the SETUP window. The second mouse click on the model name automatically closes the MODEL window and opens the SETUP window.

A.3.2 SETUP selection window

Once a specific model is activated, it usually is appropriate to select a preprogrammed model configuration with the SETUP window (fig. A.3-2). The list of setup options included in this window provides a variety of specific model configurations. Each of these configurations includes the specifications for which windows are open initially, the variables that are to be graphed and the parameter values. The complete list of SETUP names is viewed either by an appropriately sized window or by scrolling. To select a specific configuration, highlight a name in the SETUP window (by clicking the mouse) and then click

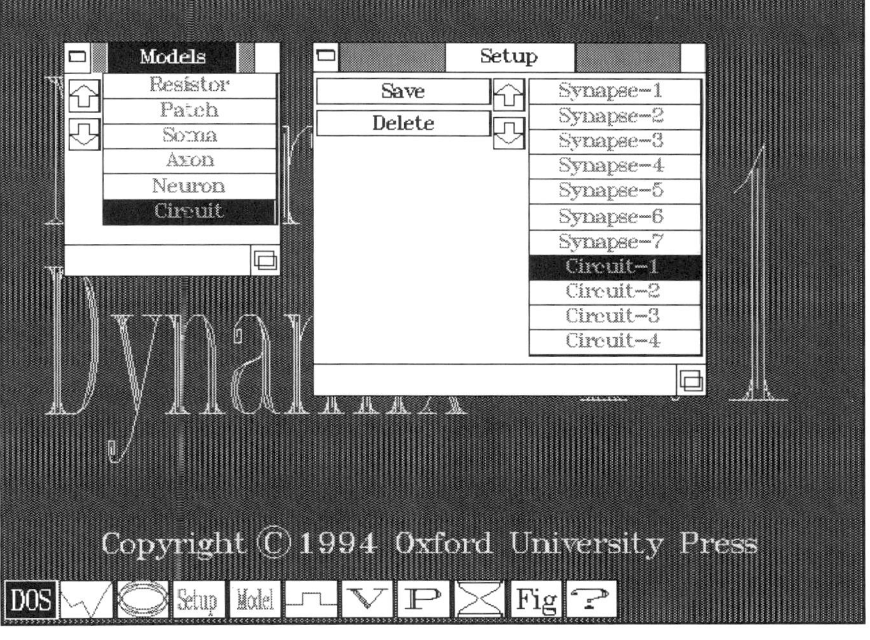

Figure A.3-1. Monitor display of the MODEL and SETUP windows

the mouse again to activate the selected setup. When all of the appropriate windows are in place on the screen, begin the modeling experiment by clicking the mouse on the STOP/GO icon (set the icon to "GO") in the TIMESERIES or PHASEPLANE windows (or alter the configuration to suit your specific modeling requirements).

To save the current configuration of *NeuroDynamix*, generate a new name for the setup by clicking the mouse on "Save" in the SETUP window. The word "Save" is now replaced with "—." Type in a new name from the keyboard; then press "enter." The current configuration is now saved in the parameter file associated with the active model and the new name is appended to the list of setups in the SETUP window. A previously saved setup configuration can be deleted by highlighting the name and then clicking the mouse on "Delete." Click the mouse a second time to confirm the deletion. Note that the names in the SETUP window are internal to PARAMETER files; they are NOT themselves file names. The name of the current PARAMETER file, which includes all of the currently available CONFIGURATIONS, is given at the top of the SETUP window. The extension PRM is common to all setup disk files. Examples of setup file names are PATCH.PRM and SOMA.PRM. If desired, old setup files can be saved by copying them to a new file named appropriately, for example, as SOMAPRM.OLD. This parameter file can then be accessed for modeling experiments by copying it to a file named SOMA.PRM.

A.3.3 TIMESERIES window

This window provides a visual display of model variables graphed as a function of time. The currently active setup for the current model is listed at the upper border (fig. A.3-2, left). The ordinate for TIMESERIES graphs represents amplitude (membrane potential, current, conductance—in appropriate units), and the abscissa represents time in milliseconds or seconds. (See section II for the specific units associated with each model variable.) The appearance of the graphs and the execution of the models are controlled by four icons at the left margin and nine icons in the lower border as described later.

A.3.3.1 Ordinate scaling The size of the amplitude excursions for graphed variables is controlled by clicking the mouse on the DILATION and CONTRACTION icons at the left of the window (double arrows). Clicking on the single UP or DOWN arrow icon offsets the display to higher or lower values. These changes in the ordinate are immediately reflected by the relabeling of the axis.

A.3.3.2. Abscissa scaling The time axis (abscissa) is rescaled by clicking on one of the two double arrow (DILATION and CONTRACTION) icons near the right edge of the lower window border. These changes in the time axis are immediately reflected by relabeling. For rapid changes in scale, click the mouse twice for each scaling step. The units of the axis represent time, as indicated in the box in the lower window border.

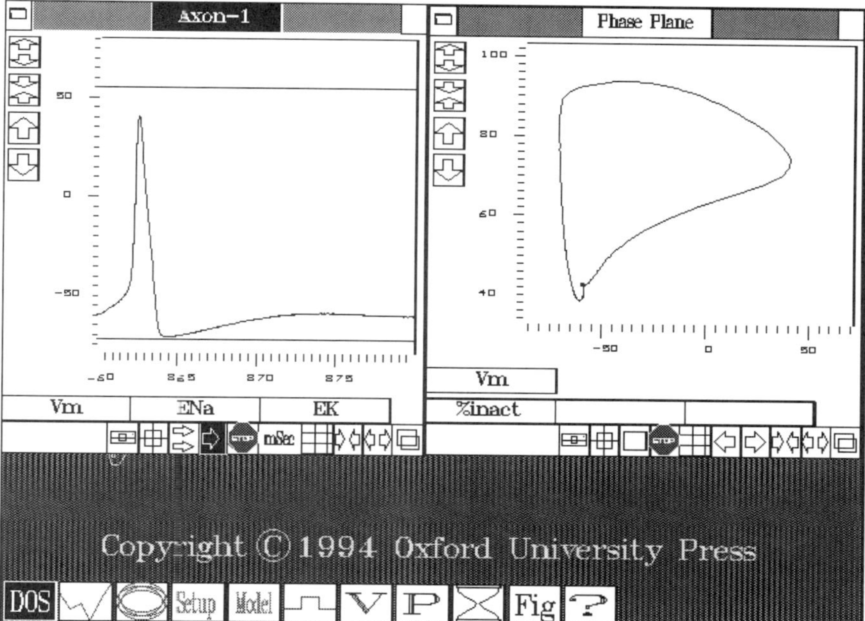

Figure A.3-2. Monitor display of the TIMESERIES (*left*) and PHASEPLANE (*right*) windows

A.3.3.3 Starting/stopping Graphing is initiated by clicking the mouse on the STOP/GO icon. This is a two-state icon that reads STOP when graphing (and calculation) is halted and GO when graphing is in progress. This control icon is also found on the CLOCK window. (Click the mouse twice on this icon if graphing does not commence even though the icon displays GO.)

A.3.3.4 Displaying a grid overlay The graph window is overlaid with a grid pattern when the mouse is clicked on the GRID icon (next to the "time" box). This icon serves as a toggle to display or remove the grid lines. (Please note that the window will be redrawn when the grid pattern is altered; as a consequence, the displayed graph may be erased.)

A.3.3.5 Pause after graphing a full window Graphing (and calculation) will pause when the traces reach the right edge of the graph display area if the PAUSE icon is highlighted (activated by clicking the mouse on the single arrow and vertical line icon). This PAUSE toggle allows inspection of the displayed graphs. Graphing is resumed when the mouse is clicked on the STOP/GO icon.

A.3.3.6 Superimposition of graph traces For some modeling experiments (e.g., AXON voltage-clamp experiments) it is useful to superimpose successive traces by not erasing previous traces each time a new sweep is begun. Superimposition is accomplished by clicking the mouse on the SUPERIMPOSITION icon (double arrows). When the SUPERIMPOSITION icon is highlighted, the graph window is not erased when the traces begin a new sweep at the left of the window. The normal graph mode, with erasure after every sweep, is re-established by clicking the mouse a second time on the SUPERIMPOSITION icon.

A.3.3.7 Saving graphs The *x* and *y* coordinates of graphed information are saved to ASCII files when the mouse is clicked on the DISK icon (left-most icon in the lower border). All data graphed in the TIMESERIES window are saved as pairs of values, with a different ASCII file for each variable. Data is directed to these files for as long as the DISK icon is highlighted. Data storage is terminated by clicking on the DISK icon a second time. The graphed points are stored in the *NeuroDynamix* directory with file names that are formed according to the following syntax: (!)(LETTER)(VARIABLE NAME)(.) (MONTH)(DAY). For example, the graph points for potassium current (IK) would be saved on February 15 as *!AIK.215*. If more than one graph of a given variable is saved on a given day, the letter in the second position of the file name is incremented for each additional data set saved.

A.3.3.8 Analyzing graphs The values of the *x* and *y* coordinates may be printed on the screen for any point in the TIMESERIES window by clicking the mouse on the CURSOR icon (cross hairs, fig. A.3-3). This causes the display of a cross-hairs cursor and opens the ANALYZE window. (You may need to move the ANALYZE window so that it does not overlap with the TIMESERIES window.) Click the mouse inside the square near the center of the TIMESERIES window to activate the cross hairs. The intersection of the cross hairs defines a point whose current position is given in the ANALYZE window next to "X-value" and "Y-value." Differences in the coordinates of any two points are

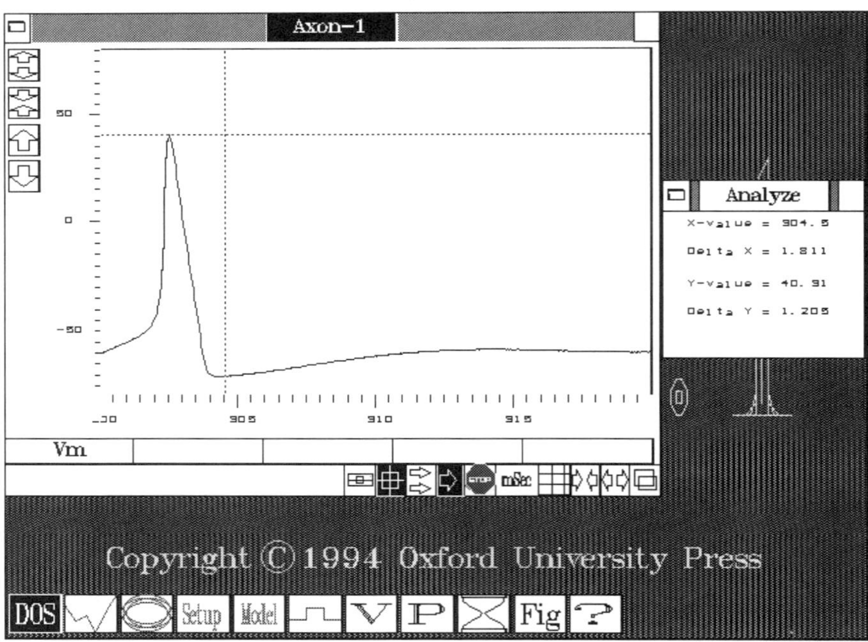

Figure A.3-3. Monitor display of the TIMESERIES (*left*) and ANALYZE (*right*) windows

obtained as follows. The mouse is clicked with the cross hairs centered at the first point (double click the first time this feature is used); then the cross hairs are moved (via the mouse) to the second point and the mouse is clicked again. Differences in ordinate and abscissa values of the two points are now displayed in the ANALYZE window next to the labels "Delta X" and "Delta Y." To exit from the ANALYZE window click RIGHT and LEFT mouse buttons sequentially (or click the middle button on a 3-button mouse); then click the mouse on the CLOSE icon of the ANALYZE window.

A.3.4 PHASEPLANE window

This window provides a visual display of model variables graphed parametrically (time is the parameter) against each other (fig. A.3-2, right). The name PHASEPLANE is given in the top border. The abscissa for PHASEPLANE graphs represents the amplitude of the first selected variable (upper name at the lower-left of the window) and the ordinate represents other variables (lower names; more than one may be selected). Because time does not appear in this window, graphing rate depends on the time scale selected in a TIMESERIES window. The appearance of the graphs and the execution of the models are controlled by icons at the left margin and in the lower border as described for the TIMESERIES window. The PHASEPLANE window includes an ERASE icon (open square in lower border) to clear this window as desired.

A.3.4.1 Ordinate scaling Same as for the TIMESERIES window.

A.3.4.2 Abscissa scaling Same as for the ordinate. Most of the controls found in the TIMESERIES window are found here as well.

A.3.4.3 Saving graphs The x and y coordinates of graphed information are saved to ASCII files when the mouse is clicked on the DISK icon (left-most icon in the bottom border). All data graphed in the PHASEPLANE window are saved in an ASCII file as columns of values beginning with the x-variable, followed by the y-variable. Data is directed to this file for as long as the DISK icon is highlighted. Data storage is terminated by clicking the mouse on the DISK icon a second time. The graphed points are stored in the *NeuroDynamix* directory file with names that are formed according to the following syntax: (!)(PHASPLN(.)(MONTH)(DAY). For example, the graph points for the PHASEPLANE would be saved on February 15 as *!PHASPLN.215*. Time series files, with data formatted as described earlier for the TIMESERIES window, also are generated simultaneously. The phaseplane data file name, unlike those for the time series data, is not incremented. **Hence, if more than one graph of the PHASEPLANE is saved on a given day, the first data file will be overwritten by the second one.**

A.3.5 VARIABLES window

This window provides the list of variables that can be selected for graphing (fig. A.3-4, left). The full list of available variables can be observed either by

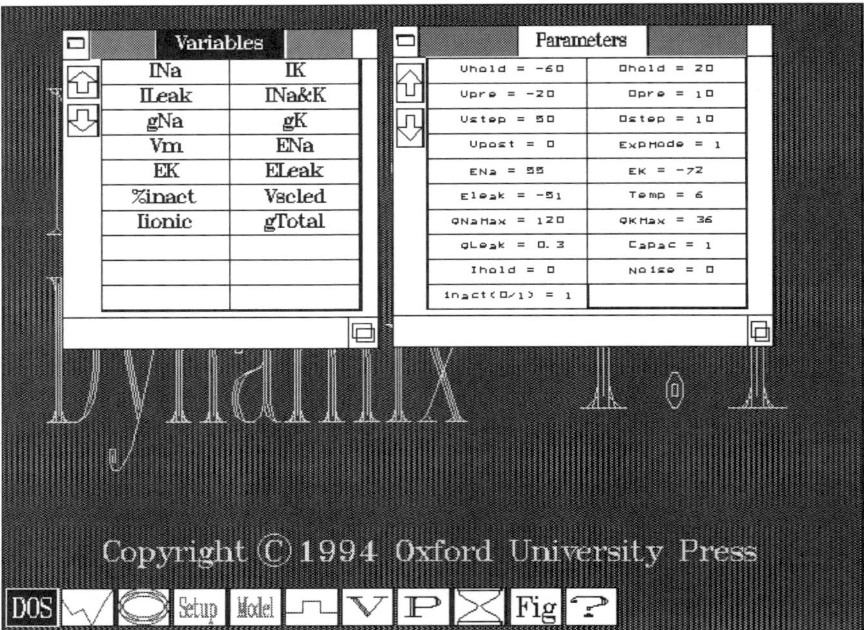

Figure A.3-4. Monitor display of the VARIABLES (*left*) and PARAMETERS (*right*) windows

enlarging the default window or by scrolling through the list. To select a variable for graphing depress and hold the mouse on a variable name and drag the rectangle surrounding the name inside the TIMESERIES or PHASEPLANE window. The name of the selected variable will now appear near the lower border of the graph window. Any number of variables may be selected for simultaneous display. Graphing of any variable is discontinued by depressing the mouse on the variable name in the window and dragging the rectangle outline outside the window. Note that the color of each variable trace corresponds to the color of the variable name at the bottom of the TIMESERIES and PHASEPLANE windows. Once the variables have been selected for graphing, the VARIABLES window should be closed. Note that if the VARIABLES window overlaps the PHASEPLANE window, the names of variables dropped into the PHASEPLANE window will not appear until the VARIABLES window is moved or closed.

A.3.6 PARAMETERS window

This window includes a list of parameters that can be altered to allow for user control of the models (fig. A.3-4, right). The full list of available parameters can be observed either by enlarging the default window or by scrolling through the list. Parameter values may be altered in two ways. Small changes can be made very quickly; simply click the RIGHT mouse button on a parameter to increment its value or click the LEFT mouse button on a parameter to decrement its value. To make larger changes in parameter values, click LEFT and RIGHT mouse buttons (middle button on a 3-button mouse) sequentially on a parameter, thereby opening the PARAMETER MODIFICATION window. This window displays the name of the parameter selected in the upper border. Below the name are a series of small boxes with the current value of the parameter. Each of the digits in these boxes can be incremented or decremented by clicking with the RIGHT and LEFT mouse buttons, respectively. The sign of the parameter can be reversed by clicking the mouse on the "+" or "−" symbol at the left. A specific parameter value is entered from the keyboard by first clicking the mouse on the "−" symbol at the bottom, entering the new value from the keyboard and then pressing the "enter" key. The new parameter value is immediately incorporated into the calculations although it may not be displayed in the PARAMETERS window until the parameter name is redrawn. Note that because of internal, program constraints, parameter values outside a specified range cannot be selected.

A.3.7 STIMULATOR window

This window controls a stimulator that provides selected waveforms to the models (fig. A.3-5). The stimulator is quite versatile, with five different output waveforms that are chosen by clicking the mouse on the icons near the upper left of the MAXIMIZED window. These waveforms are (icons, left to right): DC (continuous), SQUARE (square-wave pulse), TRIANGLE (triangular-shaped output), RAMP (sawtooth: slowly rising, instantaneous return to 0), and SINE (sine-wave output). The currently selected waveform is highlighted. The amplitude of the waveform, the duration of the output (except for DC), and the period of the output pattern (except for DC) are controlled via the three

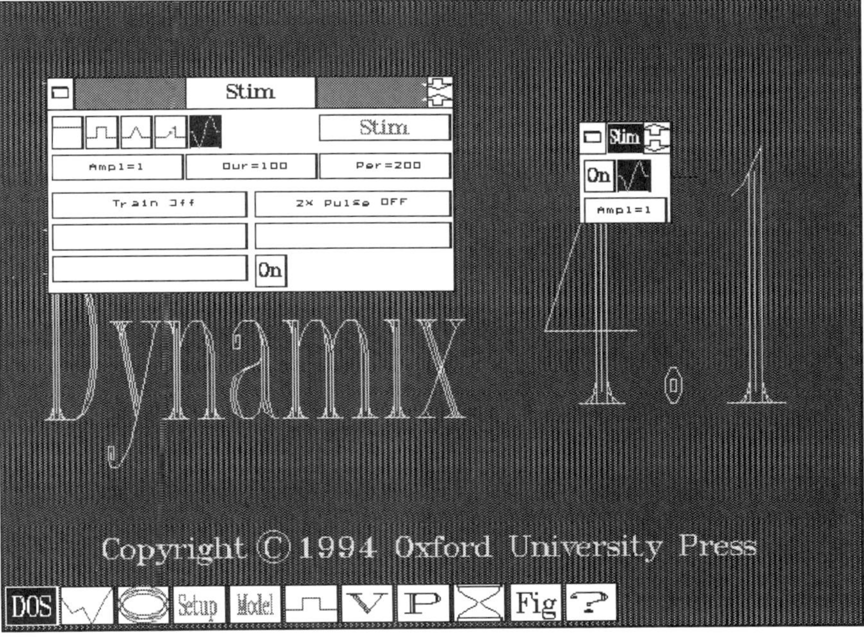

Figure A.3-5. Monitor display of the STIMULATOR window, MAXIMIZED (*left*) and MINIMIZED (*right*)

rectangles below the waveform control. Amplitude or *Ampl* (in appropriate units) refers to the peak value of the stimulator output. Duration or *Dur* (units are 1/1000 of the primary time unit for the model currently active) is the temporal extent of the waveform, from the beginning to the end of the pattern. The period or *Per* is the interval between repeated waveforms. These, and all other stimulus parameters, are selected by procedures that are identical to those used to select parameter values.

In addition to single pulses and continuously repeated waveforms, the stimulator can also generate double pulses of any of the selected waveforms (click mouse on *2x Pulse*). The delay of the second pulse is selected with *2x Pulse Delay* (same units as *Dur*). The stimulator also will generate trains of pulses (*Train on*) with selected train duration (*Train Dur*) and train period (*Train Period*).

The stimulator output is toggled on or off by clicking the mouse on the ON/OFF icon at the bottom middle of the STIMULATOR window. Output of the stimulator is graphed in the TIMESERIES and PHASEPLANE windows by dragging the rectangle around the *Stim* name (upper right) into the TIME-SERIES window.

The STIMULATOR window is rather large, but it can be reduced in size by clicking the mouse on the MINIMIZE icon (upper right-hand corner). This procedure closes the large STIMULATOR window and opens a much smaller one (fig. A.3-5, right). The reduced stimulator window includes the ON/OFF icon, a WAVEFORM icon (waveforms are selected by repeated clicking of the mouse), and a rectangle to control the amplitude of the stimulator output. The default stimulator window is reopened by clicking the mouse on the MAXIMIZE icon (upper right-hand corner).

A.3.8 TIME window

This window features a digital clock that displays running model time (fig. A.3-6, left). It also includes a RESET icon (0). Clicking the mouse on this icon resets the model clock to 0, resets the left edge of the abscissa in the TIMESERIES window to 0, and clears the TIMESERIES and PHASEPLANE windows. The SKIP icon can be clicked to increment or decrement the number of points that are to be skipped (calculated but not plotted), to achieve the aim of maximum graphing rate without undue distortion of the plots. The default value is 1 (every second point is graphed). The maximum skip value is 9 (1 point in every 10 is graphed). Note that the skipped points are not written to the disk file when data is saved. This window also has a STOP/GO icon to begin or stop graphing. Because the modeling rate is considerably slowed when this window is displayed, it should remain closed during most modeling experiments.

A.3.9 FIGURE window

This window provides a graphical display that illustrates some aspects of the specific model selected (fig. A.3-6, right). Although it can be moved to any convenient location on the screen, this window, unlike most of the others,

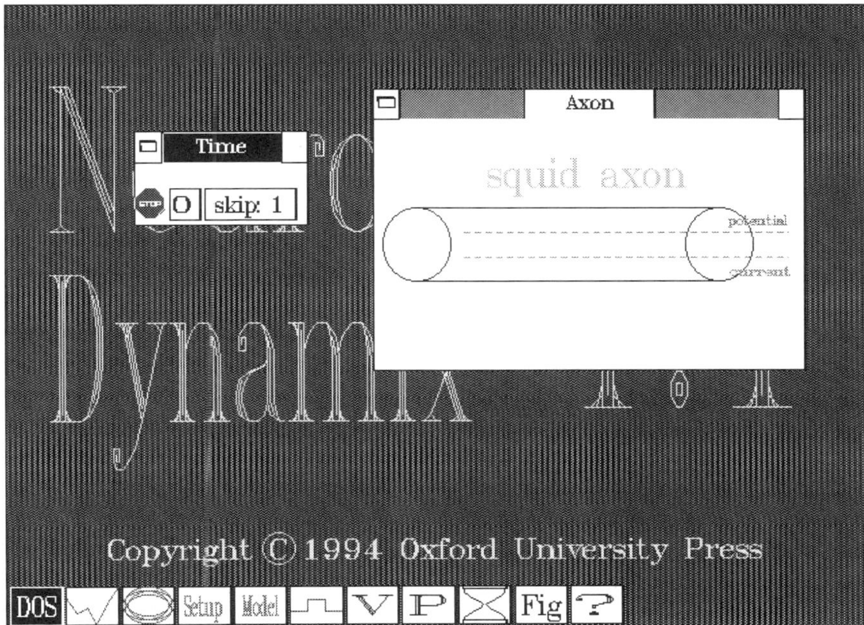

Figure A.3-6. Monitor display of the TIME (*left*) and FIGURE (*right*) windows

cannot be resized. For some models the window display is interactive, with the illustration reflecting the specific selected model parameters.

A.3.10 HELP window

The HELP window provides on-line help for using *NeuroDynamix* windows. The window displays the same information that is found in the appendix to the text. For help on a specific topic first open the HELP window by clicking the mouse on the HELP [?] icon; then click again on the appropriate heading on the list displayed in the window. You may scan the displayed text either by scrolling (click the LEFT mouse button on an arrow in the left window margin) or by paging (click the RIGHT mouse button on an arrow in the left window; fig. A.3-7). Note that the help information is not context sensitive—all of the help information is constantly accessible during the program execution by clicking the mouse on the HELP [?] icon at the lower border of the main screen.

A.4 QUITTING *NEURODYNAMIX*

Modeling sessions are terminated by clicking the mouse on the DOS icon or by pressing "q" (lowercase) on the keyboard. If you have established a new parameter set or model configuration, you may wish to save the current configuration of *NeuroDynamix* with the help of the SETUP window.

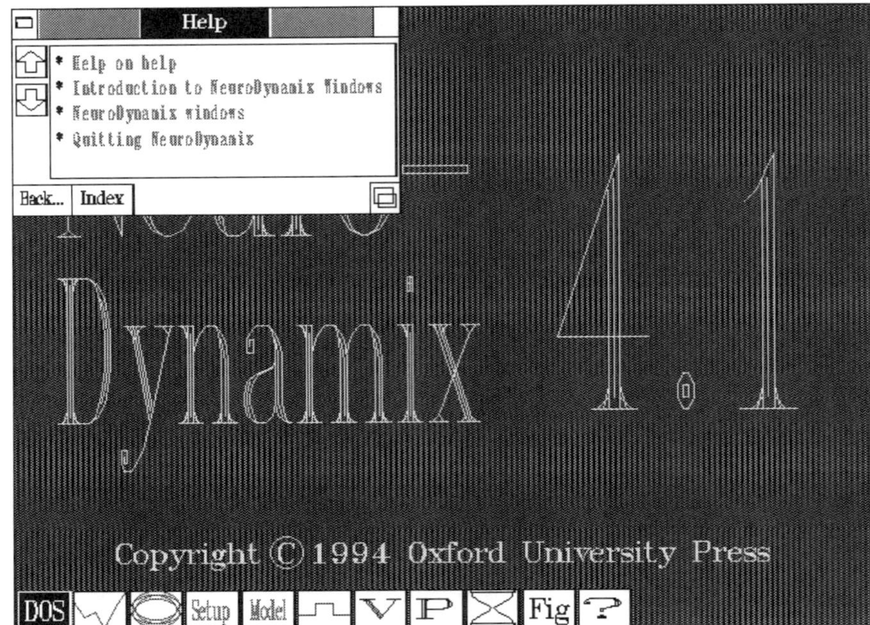

Figure A.3-7. Monitor display of HELP window

Quick Reference Guide

(RIGHT/LEFT refers to right/left mouse buttons)
(BOTH—click both buttons on a 2-button mouse or the middle button on a 3-button mouse)

Initial icons: DOS, TIMESERIES, PHASEPLANE, SETUP, MODEL, STIMULATOR, VARIABLES, PARAMETERS, CLOCK, FIGURE, HELP

DOS Return to disk operating system, also *"q"*

WINDOWS (all): Upper left icon: close window
Top border: move window while depressing LEFT
Lower right icon: resize window while depressing LEFT

TIMESERIES and PHASEPLANE (TS and PP)

Icons (left): *Dilate scale, contract scale, offset up, offset down*

Icons (bottom): *Disk*: save data to disk while highlighted

Analyze: turn on cross hairs, display ANALYZE window

Multiple traces: graph without erasing previous graph (TS only)

Stop at right: pause graphing when trace is at right edge (TS only)

Erase: erase the graph (PP only)

Stop/go: pause/commence graphing (and calculation)

Sec/msec: indicates units for abscissa (time) (TS only)

Grid: turn grid on/off

Contract time scale: use if graphed points extend beyond window

Dilate time scale: use if graphed points do not fill window

Reset variables: click on screen, position cross hairs click again

Variables: All variables are graphed as functions of time (TS only); first variable is x-variable, others are y-variables (PP only); LEFT scrolls one line, RIGHT scrolls one page

SETUP Select, save, and delete SETUP options

MODEL RESISTOR, PATCH, SOMA, AXON, NEURON, CIRCUIT

STIMULATOR Waveform select; amplitude; duration; train (on/off and select parameters); double pulses (on/off and select interval); graph stimulus value (drop name on TIME-SERIES); reduce window size (RIGHT on upper right icon)

VARIABLES Select by dropping name outlines onto TIMESERIES and PHASEPLANE windows; deselect by moving name outline from window

PARAMETERS Increase values: RIGHT; decease values: LEFT; PARAMETER MODIFICATION window: BOTH (change digits with RIGHT/LEFT, or LEFT on "−", type in number and "enter"); LEFT scrolls one line, RIGHT scrolls one page

TIME Reset time; view elapsed time; STOP/GO; SKIP points (0–9)

Bibliography

INTRODUCTORY NEUROBIOLOGY TEXTS

Aidley, D. J. (1989). *The Physiology of Excitable Cells*. New York: Cambridge University Press.

Kandel, E. R., J. H. Schwartz, and T. M. Jessell (1991). *Principles of Neural Science*. 3rd ed. New York: Elsevier.

Nicholls, J. G., A. R. Martin, and B. G. Wallace (1992). *From Neuron to Brain*. 3rd ed. Sunderland, MA: Sinauer.

Shepherd, G. M. (1988). *Neurobiology*. 2nd ed. New York: Oxford University Press.

ADVANCED NEUROBIOLOGY TEXTS

Hille, B. (1992). *Ionic Channels of Excitable Membranes*. Sunderland, MA: Sinauer.

Junge, D. (1992). *Nerve and Muscle Excitation*. Sunderland, MA: Sinauer.

SPECIALIZED REFERENCES

Membrane channels and patch-clamp recording

Catterall, W. A. (1988). Structure and function of voltage-sensitive ion channels. *Science* 242:50–61.

Neher, E. (1992). Ion channels for communication between and within cells. *Science* 256:498-502.

Sakmann, B. (1992). Elementary steps in synaptic transmission revealed by currents through single ion channels. *Science* 256:503–12.

Origin of the membrane resting potential

Bernstein, J. (1902). Untersuchungen zur Thermodynamik der bioelektrischen Ströme. *Pflügers Arch.* 82:521–62.

Hodgkin, A. L., and P. Horowicz (1959). The influence of potassium and chloride ions on the membrane potential of single muscle fibres. *J. Physiol.* 148:127–60.

Livengood, D., and K. Kusano (1972). Evidence for an electrogenic sodium pump in follower cells of the lobster cardiac ganglion. *J. Neurophysiol.* 35:170–86.

Basis of the nerve impulse

Curtis, H. J., and K. S. Cole (1939). Electric impedance of the squid giant axon during activity. *J. Gen. Physiol.* 22:649–70.

Hodgkin, A. L., and A. F. Huxley (1952a). Currents carried by sodium and potassium ions through the membrane of the giant axon of *Loligo*. *J. Physiol.* (London) 116:449–72.

Hodgkin, A. L., and A. F. Huxley (1952b). The components of membrane conductance in the membrane of the giant axon of *Loligo*. *J. Physiol.* (London) 116:473–96.

Hodgkin, A. L., and A. F. Huxley (1952c). The dual effect of membrane potential on sodium conductance in the giant axon of *Loligo*. *J. Physiol.* (London) 116:497–506.

Hodgkin, A. L., and A. F. Huxley (1952d). A quantitative description of membrane current and its application to conduction and excitation in nerve. *J. Physiol.* (London) 116:500–44.

Hodgkin, A. L., and B. Katz (1949). The effect of sodium ions on the electrical activity of the giant axon of the squid. *J. Physiol.* (London) 108:37–77.

Hodgkin, A. L., A. F. Huxley, and B. Katz (1952). Measurements of the current-voltage relations in the membrane of the giant axon of *Loligo*. *J. Physiol.* (London) 116:424–48.

Synaptic transmission

Dale, H. H., W. Feldberg, and M. Vogt (1936). Release of acetylcholine at voluntary motor nerve endings. *J. Physiol.* 86:353–80.

Fatt, P., and B Katz (1951). An analysis of the end-plate potential recorded with an intracellular electrode. *J. Physiol.* 115:320–70.

Furshpan, E. J., and D. D. Potter (1959). Transmission at the giant motor synapses of crayfish. *J. Physiol.* 145:289–325.

Loewi, O. (1921). Uber humorale Übertragbarkeit der Herznervenwirkung. *Pflügers Arch.* 189:239–42.

Oscillations in neuronal circuits

Adrian, E. D. (1931). Potential changes in the isolated nervous system of *Dytiscus* marginalis. *J. Physiol.* 72:132–51.

Angstadt, J. D., and R. L. Calabrese (1989). A hyperpolarization-activated inward current in heart interneurons of the medicinal leech. *J. Neurosci.* 9:2846–57.

Calabrese, R. L., and E. A. Arbas (1989). Central and peripheral oscillators generating heartbeat in the leech *Hirudo medicinalis*. In *Cellular and Neuronal Oscillators*. (J. W. Jacklet, ed.). New York: Marcel Dekker, pp. 237–63.

Friesen, W. O. (1989). Neuronal control of leech swimming movements. In *Cellular and Neuronal Oscillators*. (J. W. Jacklet, ed.). New York: Marcel Dekker, pp. 269–316.

Getting, P. A. (1989). A network oscillator underlying swimming in *Tritonia*. In *Cellular and Neuronal Oscillators*. (J. W. Jacklet, ed.). New York: Marcel Dekker, pp. 215–36.

Perkel, D. H., and B. Mulloney (1974). Motor pattern production in reciprocally inhibitory neurons exhibiting postinhibitory rebound. *Science* 185:181–83.

Stent, G. S., W. J. Thompson, and R. L. Calabrese (1979). Neural control of heartbeat in the leech and in some other invertebrates. *Physiological Reviews*. 59(1):101–36.

Szèkely, G. (1965). Logical network for controlling limb movement in urodela. *Acta Physiol. Acad. Sci. Hung*. 27:285–89.

Willows, A. O. D., D. A. Dorsett, and G. Hoyle (1973). The neuronal basis of behavior in *Tritonia*. II. Neuronal mechanisms of a fixed action pattern. *J. Neurobiol.* 4:255–85.

Wilson, D. M., and I. Waldron (1968). Models for the generation of motor output pattern in flying locusts. *Proc. Inst. Elec. Electron. Engrs*. 56:1058–64.

NeuroDynamix 4.1
Installation and Configuration Instructions

Please make a backup copy of your program disk before beginning the installation process.

I. Installing *NeuroDynamix* on your fixed disk

To install this neuronal simulation program manually, first make a directory on your fixed disk named "ND." Then copy all of the files on the 3-1/2″ program diskette into the ND directory of your fixed disk. Now select this directory; type "ND"; and press the ENTER key. You may use the install batch file on the NeuroDynamix program diskette to carry out this procedure for you. Simply put the diskette into your floppy drive, select that drive, and then type "INSTALL." The NeuroDynamix files will now be placed into a directory named "ND" on your C: fixed disk.

II. Configuring *NeuroDynamix* for your computer

NeuroDynamix is designed to run on DOS-based PC computers. The minimum configuration requires an EGA graphics card, a math coprocessor (included on IBM 486DX compatible machines), and a 2- or 3-button mouse. The default configuration when you install *NeuroDynamix* includes VGA graphics and a 2-button mouse. If your system has EGA graphics or a 3-button mouse, you can reconfigure *NeuroDynamix* by selecting the ND directory and typing "NDSETUP." Scroll down the setup menu with your cursor keys to select the "Graphics Card" option; press the ENTER key to obtain a list of graphics options. Highlight the appropriate option for your system and press ENTER. If your mouse has three buttons, move the cursor to the "Mouse Buttons" option and press ENTER. Type a "3" in the highlighted space and press ENTER. At this point you may also wish to change the colors of the initial monitor screen. The setup program has some additional, advanced customizing features that you may choose to ignore. Scroll down to the bottom of the options list to highlight "Save configuration" to save your new setup configuration and return to DOS. Type "ND" and press the ENTER key to start *NeuroDynamix*.

III. Minimizing the disk space occupied by *NeuroDynamix* files

If space on your fixed disk is limited, remove unnecessary files after program installation by running the batch file NDMIN.BAT; simply type "NDMIN" and press the ENTER key. This will remove all text and installation files from the C:\ND directory on your fixed disk. **Please note that this will also remove the on-line help information.**